Studies in International Economics and Institutions

Studies in International Economics and Institutions

H.-J. Vosgerau (Ed.), New Institutional Arrangements for the World Economy
IX, 482 pages. 1989

M. Rauscher, OPEC and the Price of Petroleum
XII, 206 pages. 1989

F. Gehrels, H. Herberg, H. Schneider, H.-J. Vosgerau (Eds.), Real Adjustment Processes under Floating Exchange Rates
VI, 302 pages. 1990

T. Tivig, Flexible Wechselkurse aus der Sicht des Finanzmarktansatzes
IX, 226 Seiten. 1991

S. Berninghaus, H. G. Seifert-Vogt, International Migration Under Incomplete Information
VIII, 116 pages. 1991

H. Kräger, K. F. Zimmermann (Eds.), Export Activity and Strategic Trade Policy
VI, 190 pages. 1992

Jesko Hentschel

Imports and Growth in Highly Indebted Countries

An Empirical Study

With 10 Figures and 44 Tables

Springer-Verlag
Berlin Heidelberg New York
London Paris Tokyo
Hong Kong Barcelona
Budapest

JESKO HENTSCHEL
Zentrum für Internationale Wirtschaft
Universität Konstanz
Postfach 55 60
D-7750 Konstanz, FRG

Korrespondenzadresse:

Von-Anckeln-Str. 4
D-2050 Hamburg 80, FRG

ISBN-13: 978-3-642-46772-1 e-ISBN-13: 978-3-642-46770-7
DOI: 10.1007/978-3-642-46770-7

Softcover reprint of the hardcover 1st edition 1992

2142/7130-543210 - Printed on acid-free paper

für meine Eltern

Acknowledgements

This study could not have been completed without the generous help and advice of many friends, colleagues, and mentors at the University of Konstanz, the United Nations Conference on Trade and Development, and the Kiel Institute of World Economics. It is not possible to mention all of them, but I would like to express special thanks to Prof. Dr. Horst Siebert and Prof. Dr. Gerd Ronning for supervising the study. Further, I am indebted to Dr. Axel Behrens, Christoph John, Georg Kell, Prof. Dr. Joachim Möller, Dr. Willi Nagl, and Dr. Michael Rauscher for their ongoing support and constructive criticism. Nevertheless, I am solely responsible for all remaining errors.

I would also like to thank Prof. Dr. Dr. Carsten Thomas Ebenroth for the freedom I enjoyed and the resources made available to me at the Centre on International Economics in Konstanz.

Contents

List of tables

List of figures

1. Introduction

1.1. The debt crisis, imports, and growth

After the outbreak of the debt crisis in 1982, highly indebted developing countries faced tight international capital markets. Latin American countries, in particular, found that they could no longer simply continue their previous policy of rolling over debt by financing amortization and interest payments on old debt with new loans from commercial banks. They had no option but to enter into rounds of rescheduling with both official and commercial creditors.

Most heavily indebted countries recorded a trade surplus in the years following the outbreak of the debt crisis and were consequently in a position to finance a net resource transfer abroad. As most of them had recorded high trade deficits in the years leading up to the disruption, this implied a major activation of the trade account. However, a rise in export revenues generally contributed little to the trade account amelioration. On the contrary, export receipts often stagnated or even declined. The burden of adjustment consequently fell on imports. In some Latin American countries like Argentina, Chile, Mexico, or Uruguay, nominal and real imports dropped by more than fifty percent.

Quite a few studies have examined the potential improvements in the foreign trade accounts of debtor countries by evaluating the scope for export expansions and import reductions. One branch of this literature incorporates standard import functions, in which imports are treated as consumer goods. But the composition of trade of the developing economies invalidates the characterization of imports as consumption goods only. The bulk of imports of these countries consists of intermediate and capital goods which enter into the production process. Consumer goods account frequently for only 10% of the import bill. The structural relationship between imported inputs and domestic output is not taken into account by these models. Import reductions, which fell largely on investment goods and intermediate supplies, do not have an effect on total output in such frameworks.

Another type of studies treats the supply-side linkage between imports and output explicitly. But they often postulate the other extreme relationship and assume an inflexible link between the two variables. The import-intensity of output growth is characterized as rigid, not influencable by, e.g., relative price changes. Hence, import reductions inevitably lead to an output shortfall.

At the heart of the different approaches are thus conflicting assumptions about the degree to which domestic capital and labor can replace imported intermediate and investment goods in developing countries. If imports are not treated as a separate factor of production, it is assumed that they can be perfectly substituted for by domestic factors. On the other hand, if imports always stand in a fixed relation to output, they are described as perfect complements to other production means.

The degree of substitutability between imported and domestic factors of production is not a merely technical or academic issue but has far-reaching policy implications. First, it is an important variable linking the developing economy to the international environment. The less substitutability for imported inputs exists, the more will external shocks like an increase in import prices or a rise in international interest rates influence domestic economic performance. For example, if imports consist largely of oil products, a rise in import prices of oil depresses oil import volumes at a given amount of exports and a balanced trade account. If domestic substitutes for the imported oil (i.e. other energy sources) are lacking, this has a negative impact on domestic output.

Second, it touches on the design and effectiveness of adjustment policies to balance of payments problems. The more complementary imported factors of production are towards domestic ones, the less effective are currency devaluations to curb the amount of imports while holding domestic output fixed. The effect of the devaluation on the trade balance consequently varies with the degree of substitutability. The trade balance is usually assumed to be a function of relative prices and gross national product. If the relative price mechanism is ineffective due to low substitutability on the import side and an unresponsive export supply, growth has to be slowed down in order to meet a fixed target for the trade account. Further, if imported inputs are complements and account for a substantial portion of total factor inputs, devaluations will exert a strong pressure on the domestic price level due to their supply-inflationary effect. This not only erodes the initial nominal devaluation but also causes internal stability problems.

Third, the relationship between imported and domestic production means is associated with one of the major research topics of development economics: the trade and industrial strategies of the developing economies. The adoption of inward-looking policies in a lot of developing countries was defended on the grounds that the countries wanted to achieve higher 'independence' from manufactures produced in the North. The economic costs going hand in hand with these policies (inefficient allocation of resources) were said to be balanced by the emerging ability to produce previously imported manufactures. If this is true, countries that have pursued a decade-long policy of restricting the supply of imports in order to provide emerging industries with a protective shield should command higher substitution possibilities for imported inputs than countries which have not adopted such a strategy.

This study estimates macroeconomic production functions for a number of developing countries which differ with respect to the level of economic development and trade and industrial strategies. The parameter of most interest is the elasticity of substitution between imported and domestic means of production. This allows us to answer several general questions which are associated with the role of imported capital and intermediate goods in developing economies. Are they substitutes or complements to domestic capital and labor? Hence, how important is the price mechanism for determining factor use? How does the importance of imports in the domestic production process influence the effectiveness of devaluation policies and the reaction of the trade balance to external shocks? Furthermore, knowledge of the elasticities allows us to give some more specific answers since the analysis is conducted on an individual country basis. Can developing countries be treated as a more or less homogeneous group, as is generally done in North-South models, or do they show significant differences in production structures? Are countries which have striven for 'independence' and self-sufficiency better able to absorb a foreign exchange shortage than countries not following such a strategy?

The supply-side linkages between imports and output are analyzed for twelve developing countries. They include all highly indebted ones (World Bank classification) for which the required data could be obtained (Argentina, Brazil, Chile, Columbia, Ecuador, Mexico, Peru, the Philippines, Uruguay, and Venezuela). Additionally, two countries are included which did not encounter major debt difficulties, but which had to cut real import levels due to balance of payments difficulties for some period during the 1980s. These are two ASEAN countries, namely Indonesia and Malaysia. A quick glance at Table 1.1 shows us that for most countries large import reductions were the major adjustment mode adopted to achieve trade balance improvements after 1981. We can also observe that growth picked up for some countries in 1985-1987 while imports (relating here to nominal values only) declined. This already suggests that a linear relationship between the growth of imports and growth of GDP as employed in most forecasting models of trade and growth are questionable.

Table 1.1: Change of the trade balance, imports, exports (million dollars), and average real GDP growth rates (per cent), 1982-1987[1]

Country	Year	Change in exports (1)	Change in imports (2)	Change in trade balance [(1)-(2)]	Real GDP growth
Argentina	1982-1984	-501	-3574	3073	-2.3
	1985-1987	-624	63	-687	.8
Brazil	1982-1984	3482	-4740	8222	.5
	1985-1987	1661	-2195	3856	6.2
Chile	1982-1984	-396	-2109	1713	-2.5
	1985-1987	680	68	612	4.2
Columbia	1982-1984	-76	619	-695	2.1
	1985-1987	1429	-991	2420	4.0
Ecuador	1982-1984	33	-506	539	1.8
	1985-1987	-62	40	-102	2.0
Indonesia	1982-1984	-364	4050	-4414	3.9
	1985-1987	-3019	-4402	1383	4.0
Malaysia	1982-1984	2173	3106	-933	7.1
	1985-1987	1495	-1906	3401	3.0
Mexico	1982-1984	7645	-6929	14574	2.0
	1985-1987	-3152	890	-4042	.4
Peru	1982-1984	-395	-86	-309	-.8
	1985-1987	-451	-378	-73	3.3
Philippines	1982-1984	-226	-199	-27	1.9
	1985-1987	-80	-1444	1364	-2.0
Uruguay	1982-1984	87	-639	726	-3.6
	1985-1987	-69	12	-81	1.1
Venezuela	1982-1984	-2143	-1917	-226	-2.2
	1985-1987	-4131	-1009	-3122	1.6

Source: IMF (1990a) and World Bank (1989d).

1 Figures represent changes of average imports, exports, and the trade balance compared to the three-year previous period.

1.2. Structure of the study

In order to provide a background for the study, Chapter 2 describes the role of imported factors of production in the twelve developing countries. It starts with a brief analysis of the composition of imports in order to evaluate whether factor demands are more important in determining imports than consumer demands. Then we look at capital and intermediate products separately. The nature of capital goods fabrication and industrial supplies production is described and factors are outlined which can impede perfect substitutability between imported and domestic manufactures. Trade and industrial statistics are matched, which allows us to assess the contribution of imports in apparent consumption of broad end-use categories. These figures are evaluated also with respect to the development strategy and policies which have been followed by the individual countries.

Chapter 3 reviews studies on imports as inputs into production. The main conclusions from theoretical growth models are pointed out, namely that the steady-state growth of an economy that imports a means of production varies with the degree of substitutability for this imported input. Therefore, empirical analysis should not only be concerned with assessing whether domestic substitutes for imported capital and intermediate goods are available or not, but also how large the elasticities of substitution are. The discussion then concentrates on empirical studies. The 'standard' import function, which does not explicitly treat imports as means of production, and modifications concerning, for example, foreign exchange availability, are discussed. Two approaches which both take account of the role of imported capital and intermediate goods are then contrasted: derived factor demands from constant elasticity of substitution (CES) and translog production functions. Assumptions underlying the two approaches are outlined and their applicability is evaluated.

Chapter 4 specifies the supply side theoretically. We choose a macroeconomic production function of the nested CES type to analyze the supply side relationships in developing countries. Imported and intermediate goods, labor, and two different capital stocks enter the production function. One capital stock is made from domestically fabricated machinery, the other one comprises only imported capital goods. Cost minimization is applied to the system and reduced form equations suitable for parameter estimation are derived.

Regression results of the elasticity of substitution between domestic and imported factors of production for the twelve developing countries are presented in Chapter 5. As the generation of data for both endogenous and exogenous variables requires assumptions about such parameters as the depreciation rates, we conduct a variety of different calculations in order to test the stability of the calculated elasticities. One modification is concerned with the use of a government reaction function to represent trade policies instead of relying on a protection proxy, which can only be employed under restrictive assumptions. Estimated substitution elasticities for both types of imports are discussed and factors causing them are pointed out by means of rank correlation analysis.

Chapter 6 specifies a macro-econometric model of merchandise trade which is able to represent the structural importance of different import categories. We estimate the model for two countries which differ with respect to the role of imported capital and intermediate goods for domestic output. External shocks are introduced. They include a sudden import price rise, a shift of external demand, and a nominal devaluation. Trade balance reactions to these shocks are simulated and commented on.

Finally, the last chapter summarizes the main findings of the analysis. It assesses how far inward-oriented development strategies have lived up to their primary aim of reducing foreign exchange vulnerability. Some policy implications relating to this issue conclude the discussion.

Data requirements for the study are large. Commodity specific trade data has to be aggregated in broad end-use categories (capital, intermediate, consumer goods). Similarly, industrial production and national income data are necessary, as well as information on tariff rates, several government financial statistics, etc. Most data definitions and methods of aggregation and deflation are reported in the Appendix. A few aspects of data handling that are of importance for understanding the argument are treated in the main text.

2. The importance of imported factors of production in developing countries

2.1. Introductory remarks

The assumption made by a lot of studies of developing countries that there is a rigid link between imported intermediate and capital goods on the one hand and domestic output on the other is very strong, since it characterizes these imports as being perfect complements to domestic capital goods and labor in the production process. But there are also quite a few reasons to doubt the notion that imported means of production can be substituted for perfectly by domestic goods. This will be pointed out in this chapter. Factors that can impede such a perfect substitutability are outlined and their relevance is evaluated separately for capital and intermediate goods. Quite a few of the twelve developing countries studied here have attempted to reduce their reliance on imports by deliberately fostering domestic industries behind a protective shield. The last part of the chapter discusses some questions which are concerned with the choice of industrialization and trade policies.

Before turning to these issues, the first section examines to what extent merchandise imports of the twelve developing countries enter into the production process rather than being destined for direct consumption. If we find that nearly all trade is conducted in consumer goods, as assumed in the traditional theory of international trade, there appears to be little justification for focusing on the role of imports as factors of production. In addition to assessing the composition of imports, its variation in the 1980s (when all countries looked at here endured real import declines at some point in time) is looked at.

2.2. The composition of imports

In this section, we analyze import data according to the Standard International Trade Classification (SITC). The SITC arranges all traded goods at varying degrees of aggregation. We choose the most detailed level (five-digit numeration with 1360 entries) and assign each entry position to one of the end-use categories 'consumer goods', 'capital goods', (non-energy) 'intermediate goods' and 'energy intermediates' (see Appendix C, Table C.2). Import data on the disaggregate SITC level as reported by the individual country are obtained from the United Nations *Comtrade* database. Since the real flow of goods is of interest here rather than nominal import values, each figure at the disaggregate SITC level is deflated before an aggregation into the broad end-use categories is performed. In order to obtain precise estimates of real merchandise trade, we use nine different import price deflators per country and per year. These deflators stem from the United Nations Conference on Trade and Development (UNCTAD) database on price deflators.[1]

Results of this decomposition exercise for the period 1976-1987 are presented in Table 2.1, in which a quantity index of total imports (average 1979 to 1981 = 100) is included in addition to the shares of the four end-use categories.

As can be detected from the table, consumer goods accounted for a minor share in total imports of the twelve developing countries under study here, generally constituting less than a fifth of real merchandise imports. Conversely, the combined share of the three other categories which all enter into domestic production in the importing country, made up for at least 80 and most often more than 90 per cent of all imported goods. Thus, a general characterization of imports as consumption goods is not justified.

Non-energy intermediate goods were registered as the most important import class, except for Brazil, where energy-inputs acquired the dominant role. The oil exporting countries in the sample recorded low shares of the latter category (especially Ecuador, Mexico, Peru after 1978, and Venezuela), while Brazil, the Philippines, and Uruguay were 'heavy' importers of fuel and other energy products. The role of capital goods varied, being relatively high in Ecuador and Venezuela while low in Brazil and the Philippines.

Eight of the countries shown in Table 2.1 experienced more or less large cutbacks of real imports in the period following the outbreak of the debt crisis, the largest occuring in Argentina (43%), Chile, and Uruguay (both 38%). Columbia, Indonesia, and Malaysia reported a substantial rise of average import levels in the 1982-1984 period but large decreases thereafter. While Peruvian imports increased on average slightly in the 1982-1984 period, they also leveled off subsequently.

Like the total volume of imports, their composition also varied substantially. The following general observations can be made:[2]

---

1 See UNCTAD (1986) for a description of the price deflator system.

2 For an analysis of the changing composition of imports of indebted developing countries compare Hentschel (1989), Lopez and Thomas (1988), Mirakhor and Montiel (1987), UNCTAD (1988a), and UNCTAD (1988b, pp.110-111). For an analysis of determinants of the trade composition (such as per capita income, growth, and the country size) of developing countries see McCarthy et al. (1987). They focus on the importance and finance of imported capital goods.

Table 2.1: The composition of real imports of developing countries, 1976-1987[1]

Country	Share of capital goods (per cent)	Share of energy intermediates (per cent)	Share of non-energy intermediates (per cent)	Share of consumer goods (per cent)	Index of real imports (79/81 = 100)
Argentina					
1976-1978	22.6	20.7	51.9	4.7	58.9
1979-1981	28.2	11.4	43.5	16.8	100.0
1982-1984	22.7	8.4	61.6	7.2	57.1
1985-1987	21.7	11.2	60.4	6.6	52.6
Brazil					
1976-1978	16.5	41.8	37.3	4.2	96.0
1979-1981	13.7	43.9	37.9	4.4	100.0
1982-1984	11.3	50.7	34.2	3.7	76.1
1985-1987	10.9	42.4	38.6	8.2	76.4
Chile					
1976-1978	20.9	24.3	41.6	13.2	57.2
1979-1981	24.2	19.4	34.0	22.3	100.0
1982-1984	19.7	16.5	45.1	18.8	61.5
1985-1986	23.6	20.8	44.4	11.2	59.9
Columbia					
1976-1978	26.0	6.4	57.9	9.6	67.5
1979-1981	28.5	6.9	54.8	9.7	100.0
1982-1984	29.3	7.0	55.6	8.0	118.3
1985-1987	23.6	2.2	66.4	7.8	90.3
Ecuador					
1976-1978	39.0	1.4	48.1	11.5	80.8
1979-1981	38.9	3.2	48.0	9.8	100.0
1982-1984	28.6	1.1	60.7	9.4	89.8
1985-1987	n.a.	n.a.	n.a.	n.a.	(78.6)[2]
Indonesia					
1976-1978	28.2	13.1	42.6	16.1	81.2
1979-1981	26.2	12.4	47.8	13.6	100.0
1982-1984	30.3	16.4	46.4	6.9	155.2
1985-1986	25.0	13.2	57.0	4.8	111.8
Malaysia					
1976-1978	20.0	18.3	44.4	17.3	66.3
1979-1981	21.4	13.8	49.8	14.9	100.0
1982-1984	24.7	9.7	50.7	14.9	138.4
1985-1987	20.9	9.1	53.8	16.2	123.0

Table 2.1 (cont.)

Country	Share of capital goods (per cent)	Share of energy intermediates (per cent)	Share of non-energy intermediates (per cent)	Share of consumer goods (per cent)	Index of real imports (79/81=100)
Mexico					
1976-1978	29.1	3.9	60.6	6.4	45.9
1979-1981	32.7	1.6	56.8	8.9	100.0
1982-1984	29.4	1.7	59.8	9.1	77.6
1985-1986	25.9	3.6	60.6	9.9	75.1
Peru					
1976-1978	24.2	20.3	47.5	7.9	98.6
1979-1981	31.9	1.9	50.7	15.5	100.0
1982-1984	27.6	2.1	53.5	16.7	103.8
1985-1987	n.a.	n.a.	n.a.	n.a.	(85.0)[2]
Philippines					
1976-1978	17.0	34.7	42.5	5.8	84.0
1979-1981	19.7	32.4	42.6	5.4	100.0
1982-1984	18.1	28.3	47.6	6.0	96.0
1985-1986	10.1	29.1	53.0	7.8	72.8
Uruguay					
1976-1978	17.3	39.5	36.2	7.0	68.6
1979-1981	20.8	28.5	36.6	14.1	100.0
1982-1984	17.6	33.7	37.2	11.5	62.3
1985-1987	14.2	28.9	42.3	14.6	67.8
Venezuela					
1976-1978	40.4	0.5	44.4	14.7	105.6
1979-1981	32.9	0.5	46.4	20.1	100.0
1982-1984	28.3	1.3	51.3	19.0	86.7
1985-1986	28.6	1.6	51.1	18.6	78.0

Source: Own calculations. For data sources and method of deflation and aggregation see Appendix C. The category of 'other goods' (defined in Appendix C, Table C.2) is allocated to the four import categories shown above according to the latter's relative weight.

1 The trade figures cover total imports of the developing countries (measured in 1980 prices), using them as the reporter of the trade flows. Since several of the countries do not report trade flows till 1987, end year values shown in the table differ. Shares do not always total exactly 100 on account of rounding errors.

2 Total import data from IMF (1990a), as these countries did not report continuous trade flows beyond 1984.

(a) The weight of consumption goods tended to move in line with total real imports, especially in those countries that had used the inflow of external funds in the 'debt-accumulation period' of 1979 to 1981[3] to raise consumption imports overproportionally, namely Argentina and Chile.

(b) In the presence of declining imports, capital goods appear to have born the burden of adjustment. The relative weight of capital goods in overall imports was reduced significantly in Argentina, Columbia, Ecuador, Indonesia, Mexico, the Philippines, Uruguay, and Venezuela, when these countries encountered balance of payments problems and real imports were compressed.

(c) Conversely, the portion of imports consisting of intermediates (sum of energy and non-energy inputs) shows an anticyclical pattern in relation to the total import quantity. They tended to rise in times of import reductions, while (to a lesser extent) consumption goods and (to a larger extent) capital goods imports declined overproportionally. The shift towards non-energy intermediates was particularly large in Argentina, Chile, Columbia, Ecuador, and Indonesia, while Brazil and Uruguay (both major oil importers) report an initial alteration of the import composition in favor of energy-intermediates.

Without entering too deeply into a discussion of why such a relatively homogeneous pattern of change can be observed, we can briefly sketch two reasons which are related to the nature of the different import categories:

(a) Salas (1982, p.308) describes consumption goods as being "used as a safety valve to reduce total private imports at times when policy-makers feel that intermediate and capital good imports are exerting too much pressure on the current account deficit". It is argued that in times of a foreign exchange shortage, 'non-essential' imports are subjected to strong import barriers in order to rearrange the allocation of foreign exchange towards 'essential' capital and intermediate goods. In an IMF study, Mirakhor and Montiel (1987, p.76) argue that "developing countries shifted the burden of import compression to consumer goods for fear of the negative investment and output consequences of restricting imports of capital and intermediate goods". But as can be seen in Table 2.1, consumer goods often represented a minor, often the smallest, item in the import bill, which was mostly due to a prolonged phase of import-substitution targeted at protecting domestic consumer goods industries. As drastic import cuts occurred after 1981, the potential cushion provided by consumer goods to absorb the import shortfall was relatively small.

(b) The observed shift in the commodity composition from capital to intermediate goods might be due to an attempt by the countries to allocate foreign exchange in a way that minimizes domestic output disruption in the short run. If economies rely both on imported investment

---

3 Compare UNCTAD (1985b, p.76).

goods for capital formation and on imported intermediate supplies as inputs in the production process, the negative short term impact is higher if vital intermediate goods are lacking; the impact of reduced capital formation as a consequence of lower levels of imported capital goods is felt in the medium to longer run.[4]

In order to evaluate this alleged dependency on both imported capital and intermediate goods, it is necessary to compute elasticities of substitution between domestic and imported factors of production, as will be done later in this study. The issue of the marked shift in the composition of imports is then raised again.

2.3. Factors impeding perfect import substitutability

In his comprehensive treatment of the theoretical links between imports and economic growth, Van Bochove (1982) distinguishes several reasons which cause an imperfect substitutability for imports: non-traded specific factors, technology, and adjustment costs.

2.3.1. Non-traded specific factors

The most obvious explanation why domestic production of previously imported goods might not be possible is that the latter require non-traded factors of production which are not available domestically. Land-sites which contain natural resources are examples, as well as the climate conditions necessary to grow certain agricultural crops. Imperfect substitutes for traded goods requiring non-traded factors of production might, however, exist. For example, crude oil can be substituted for by coal, solar, or wind power as a prime supplier of energy. But since the replacement induces a different technological relationship between inputs and output, they are only imperfect substitutes. Additionally, even if a non-traded factor is completely absent, some substitutability of imports can be achieved since trade takes place at different levels of completion of a product. For example, oil is imported both in crude form as well as in the form of refined petroleum products. By changing the weight of products containing oil at different completion levels, some substitutability of the bundle of imports might be possible.

---

4 UNCTAD (1988b, p.110) holds that the choice between compressing imports of capital goods (and thereby reducing capacity expansion) and lowering intermediate goods (with the consequence of lower capacity utilization and reduced short run income and employment) "has been made at the expense of investment, since import requirements per unit of investment usually are typically much greater than for other components of domestic production".

2.3.2. Technology

Technology not only consists of the hardware input into the production process, but also comprises human skills related to knowledge about the latter.[5] Technological change is caused by innovations and involves the discovery and application of new production methods. It is therefore convenient that we distinguish between the static and the dynamic aspect of technology.

In the *static* aspect, technology can be treated as an additional factor of production. Any commodity is produced with a given level of technology at a given point in time. It determines, together with factor prices, the use of other factors of production (such as labor and capital). If a country attempts to replace imports by domestic production, the success will be largely determined by its ability to provide or aquire the necessary technology, and by the substitution possibilities between technology and other means of production.

Substitution of imported commodities is relatively easy, if the required technology for production is already existent in the economy or if the country is able to acquire it without major purchasing or adjustment costs. If this is not possible, the degree of substitutability depends on the divisibility of the technological know-how. The manufacture of complex capital goods such as engines and turbines, e.g., requires very sophisticated and largely indivisible skills in design, product engineering, and manufacture. If both divisibility and dispersion of the technology are absent, the latter becomes a non-traded specific factor.

Several studies have examined the role of technology as a factor of production in directing the international exchange of goods. Hirsch (1974, 1975) finds that the relative skill-endowment of a country versus its trading partners is an important determinant of international trade patterns. The product-cycle and neo-technology approaches to international trade argue that production of a good goes through distinct stages, associated with 'new', 'growth', and 'mature' phases, in which the factor use in the production process changes over time. "The production function of New Products is characterized by high skill-intensity, i.e. by a high proportion of scientists, engineers and skilled craftsmen in the labour force, while that of Mature Products is characterized by low skill-intensity" (Hirsch 1975, p.305). It is held that the (changing) bundle of imports and exports is associated with the relative skill-intensity of the trading economy versus its partner countries at any point in time. As will be pointed out below, the technological requirements are of particular importance when examing the indigenous manufacture of capital equipment.

---

5 For broad definitions of the technology concept compare Dahlmann et al. (1987), Stewart (1977, p.1), and Strassmann (1968, p.2).

The *dynamic* aspect of technology involves its stimulating role for innovations: Technological improvements (be they related to products or processes) are often followed by increased research and development expenditures and human capital formation which themselves are a spur to new innovations. This self-inducement of technological change can act as a barrier of entry for countries which prevents them from starting indigenous production of previously imported goods, since it complicates the process of technological catch-up.

2.3.3. Adjustment costs and time

Even if non-traded specific factors are not a motive for imports and technology is obtainable both instantaneously and without cost, production might not be undertaken if adjustment costs are high relative to the economic benefit generated by it. The discounted gain of assuming indigenous production of previously imported goods must exceed the adjustment cost.[6] A high rate of discount is caused by a strong time preference for consumption today rather than tomorrow as well as by a high degree of uncertainty concerning future price developments. For example, if it cannot be foreseen whether increased internal prices of imported goods (caused e.g. by a foreign exchange shortage via a devaluation) have to be considered as temporary or permanent, this pushes up the discount rate.

Adjustment costs are low (and the adjustment time short), if manufacture of the previously imported good does not require a major recasting of already installed capital or a relocation and training of the labor force. The manufacture of tractors not formerly produced is eased, if an automobile industry already exists. Capital goods of the automobile industry can be used to produce tractors as well; workers who are skilled auto-mechanics can be employed in tractor production. But if the economy is dominated by the agricultural sector with little or no activity in the machinery and equipment producing industries, the adjustment involves high investment costs in plants, hardware, and infrastructure. It also requires a high mobility of the work force and intensive skill accumulation. While the tractors (as a perfect substitute) might be produced in the long run, the process is obviously cost- and time-consuming.

Government policies can also impede the domestic production of an otherwise perfect substitute by raising the private cost of adjustment as encountered by the entrepreneur above the social cost. A few examples might suffice to illustrate this point. Investment licensing is practiced in a large number of developing countries, which causes potential investors to engage in an expensive licence bargaining, in which efficiency considerations often play only a minor role. Low interest rate credit policies have been observed to crowd out private savings, forcing investors to

---

6 Compare Van Bochove (1982, p.58).

obtain funds at a substantially higher rate on the black market. Protectionist trade and exchange rate policies have often led to distorted factor prices and the application of inappropriate technologies in the protected sectors.[7]

To briefly summarize, imperfect substitutability for an imported product can be caused by
- non-traded factors of production,
- lack of technological skills, and
- adjustment costs and time.

We will now evaluate the importance of these three factors for capital and intermediate goods separately.

2.4. Capital goods

While the supply of savings in an economy determines the potential amount of resources that can be invested, the availability of machinery and equipment is necessary as well in order to channel these funds into productive use. Most developing countries import the largest share of their fixed instalments of machinery from the industrialized countries, where the bulk of capital goods manufacturing in the world takes place. This stems largely from the nature of production in the capital goods sector.

2.4.1. Nature of capital goods production

2.4.1.1. Technological requirements

Capital goods industries produce a wide variety of different products, ranging from simple hand-made tools to micro-chips and merchant vessels. The manufacturing sector producing capital goods is usually sub-divided into the industries producing metal products, non-electrical and electrical machinery, transport equipment, and professional and scientific instruments.[8] Most of the production processes can be described as highly technology-intensive. If all manufacturing industries are ranked by the technological requirements in the fabrication process, capital goods

---

7 Compare, e.g., Donges (1976), Donges and Riedel (1977), Little et al. (1970), and Ranis (1979).

8 See Chudnovsky et al. (1983, p.7, fn 1), Jacobsson (1988, p.14, fn 1), and UNIDO (1979, p.172). These industries correspond to the International Statistics Industrial Classification (ISIC) categories 381 to 385. The industries also produce durable consumer goods like refrigerators, TVs, passenger motor cars, and radios.

like office machines, electronic components, electrical and non-electrical machinery, telecommunications equipment, and scientific instruments appear at the top of the list.[9] One can distinguish between four different technological inputs in the manufacturing process of capital goods:[10]

(a) *Machine operating skills* include not only the technical handling of the hardware, but also the latter's maintenance and repair. Compared to other manufacturing activities, as e.g. in the food processing industries, the machine operating skills required in the capital goods sector are much broader since they involve mastery of casting, forging, machining, heat treatment, assembly, and testing.

(b) *Manufacturing technology* is composed of organizing, managing, and executing the factory operations. In contrast to the industrial supplies industries, the bulk of capital goods production takes place in highly specialized plants with strong forward and backward linkages. The managerial know-how required to direct the production in the most efficient way involves the organization of simultaneous processes such as ensuring the supply of a wide variety of intermediates by subcontracting firms and the observance of both specific quality and delivery demands by the customers.

    The engineering knowledge of the manufacturing methods needed in capital goods production (which tools to be used at what speed, what sequence of production, how to conduct quality controls etc.) can also be described as more demanding than in other industries.

(c) *Product design capacity* is of particular importance in the capital goods sector as design requirements rise with technological complexity of the products, which is on average much higher than in other industries. Product design capabilities are strongly linked to the availability of highly skilled engineers with considerable design experience. Developing countries with an emerging capital goods sector therefore rely to a high degree on outside assistance (e.g., technical license agreements).

(d) *Research and development capacity* of the capital goods industries is larger than in other industries too since the necessary R&D expenditures rise with design requirements.

These high technology demands are often held responsible for the relatively minor importance of indigenous capital goods production in most developing countries and their predominant reliance on foreign supply of capital goods from the industrialized countries (UNCTAD 1985a, p.3). It is held that especially in countries with a weak industrial base and scarcity of skilled personnel lack of technological know-how is an important impediment to starting capital goods production.

---

9 For an industry classification by broad technological categories see UNCTAD (1987a).

10 See Chudnovsky et al. (1983, pp.10-14).

2.4.1.2. Market size

Most metal-working industries producing capital goods are characterized by a manufacturing process involving consecutive stages of production and assembly. In contrast, the production of industrial supplies (iron and steel, fertilizers, paper and pulp) takes place in so-called process industries within integrated plant facilities. Production of these intermediate goods (and some capital goods like motor assembly in vertically integrated plants) takes place according to anticipated demand, while the bulk of capital goods is manufactured according to order due to distinct stages of production and the limited amount of orders. The question of the market size is of special importance in developing countries. Although demand for a variety of capital goods might already exist even if the manufacturing base is still weak, production of specialized capital products might not be viable.

In order to explore the impact of the market size on capital goods production, the United Nations Industrial Development Organization (UNIDO) has conducted a world-wide survey of the nature and location of capital goods production (UNIDO 1988b, pp.121-132). Astonishingly, the UNIDO survey finds that quite a few developing countries produce a limited amount of capital goods which are characterized by relatively high technological complexity numbers. The technological complexity indices used in the study attempt to reflect technological requirements necessary in the manufacturing of the specific capital good rather than pertaining to the use of the product itself. UNIDO (1988b, p.129) concludes that the size of the readily accessible markets can act as a barrier of entry into capital goods production.

2.4.1.3. Economies of scale and specialization

Both economies of scale and of specialization are connected with capital goods production. *Economies of scale* occur in the manufacture of standard capital equipment like tractors, motor cars, and diesel engines. Production occurs in continuous flow operations based on expected demand, where scale economies are due to the homogeneity and large scale of output.[11]

A large part of activity in the capital goods sector does not take place in such process industries but requires design and manufacture of capital goods according to specific orders, especially in the mechanical industry. *Economies of specialization* can arise within an industry due to a high degree of vertical division of labor. A vertical separation of production takes place on account of the minimum optimum scale of production, varying among the different processes involved. Subcontracting firms often supply certain parts and components to upstream plants or

11 See Chudnovsky et al. (1983, p.16).

perform specific activities (e.g., heat treatment or forging). Specialization also pertains to the final products as firms concentrate R&D expenditures and design work on a very narrow range of products which creates distinct skills and techniques.

Jacobsson (1988) stresses the important role of an increased intra-industry specialization for trade in capital goods. He finds that especially small and medium size industrialized countries rely heavily on intra-industry trade in order to reap the returns of specialization. "The growing trade dependency means that local producers cover a diminishing part of the range of products that is demanded in the local markets and that their production of this smaller range is increasingly being sold to foreign consumers" (Jacobsson 1988, p.19). The strong and growing intra-industry trade in capital goods plays an important role in the industrial strategy of developing a capital goods sector. Quite a few developing countries have attempted to gain independence from foreign supply of capital goods by deliberately fostering domestic capital goods production and restricting imports. As technological change is rapid in these industries due primarily to intensive research and design activities on specific products, a development strategy which aims at suppressing the inflow of foreign capital goods carries the danger of detaching the domestic economy from international technological progress.

2.4.2. Weight and structure of the capital goods sector

Due to the relatively high technology requirements and the limited market size, the developing countries as a group account for only a minor fraction of world activity in the capital goods industries, namely 6% (in 1985), while in the consumer non-durable industries the value added accruing in developing countries relative to world activity is considerably higher at 15.6% and in the manufacture of industrial intermediate goods at 15.3%.[12] Moreover, while these figures manifest the minor role capital goods industries of the developing countries acquire in manufacturing value added, growth in capital goods production was also slower than in the industries producing consumer goods or industrial supplies. Unlike the growth experience of the industrialized countries, consumer goods have been the leading manufacturing sector in developing economies.

On an average, machinery and equipment producing industries account only for 10 to 15% of manufacturing value added in developing economies in contrast to 30-35% in the industrialized countries. Consequently, export earnings stemming from capital goods are still in their infancy

---

12 Data from UNIDO (1988b, pp.107-108). The classification of industries follows Table C.1 in Appendix C.

(exceptions being the Republic of Korea, Brazil, and Malaysia), while they account for a quite substantial portion of the import bill, varying around 20-30%, as can be observed in Table 2.1.

We contrast the structural composition of the manufacturing sector in the developing countries to that in three industrialized economies (the Federal Republic of Germany, Japan, and the United States of America) in Table 2.2. Columns two to four in the table show the composition of manufacturing value added (MVA) with respect to a classification of industries by end-use. The general picture that emerges is that industries producing industrial intermediate goods clearly dominate the manufacturing sector in the developing countries, closely followed by the consumer goods industries. The capital goods sector's weight in total MVA is very small, especially in Chile and the Philippines, while it accounts for about 30% of total MVA in Brazil and for somewhat less than a fourth in Malaysia and Argentina.

The minor importance of capital goods industries in the developing countries is in contrast to the dominant role these industries play in the three industrialized countries included in Table 2.2, where they account for about half of MVA. The picture is further accentuated if one compares the share of GDP originating in the five manufacturing branches classified here as producing predominantly capital goods (last column). As the share of manufacturing activity in GDP of the developing countries is on an average much lower than in the developed countries (especially when compared to Germany and Japan), the diverging structure within the manufacturing sector between the countries gets even more pronounced. While in the three industrialized countries at least 10% of economic activity had its origin in the capital goods sector, most of the developing countries studied here report only a figure of 2 to 3%. Relatively high shares can be traced for Brazil (7.6%), Malaysia (5.4%), and Argentina (5.3%).

Table 2.3 carries the analysis of the industrial structure one step further and examines the distribution of value added within the capital goods sector. The number of different capital goods' categories in which domestic activity took place in 1984 is presented in the table as well.

The metal products industry (ISIC 381) is considered to be the entry industry in the capital goods sector, since it is essential for enabling maintenance and repair work.[13] These services fuel an evolutionary development towards entry into capital goods production, which involves the formation of basic metal-working skills (such as bending, forging, and welding techniques) and domestic manufacture of simple tools and standard equipment. Products manufactured in this industry (hand tools, general hardware, fixtures of primary metal) are also essential for supplying infrastructural investment such as buildings for storage. Skill requirements for activity in the metal

13 For a description of individual industries and the sequence of production activities in the capital goods sector see UNIDO (1987, Chapter 4) and UNIDO (1988b, Chapter 4).

Table 2.2 : Structure of manufacturing value added, 1985-1987

Country	Share of MVA in GDP (per cent) (1)	Share of industries producing mainly intermediate goods (per cent) (2)	Share of industries producing mainly capital goods (per cent) (3)	Share of industries producing mainly consumer goods (per cent) (4)	Share of GDP originating in the capital goods industries (per cent) (5)
Argentina	23.9	48.9	22.1	28.9	5.3
Brazil	26.1	44.3	29.1	26.6	7.6
Chile	20.6	62.9	6.1	31.0	1.3
Columbia	22.9	43.0	12.3	44.7	2.8
Ecuador	16.4	47.6	14.0	38.4	2.3
Indonesia	14.4	62.1	11.6	26.3	1.7
Malaysia	21.6	45.6	24.7	29.7	5.4
Mexico	22.2	49.7	16.7	33.7	3.7
Peru	21.3	60.0	11.9	28.2	2.5
Philippines	23.9	45.8	7.9	46.4	1.9
Uruguay	19.9	48.7	10.3	41.0	2.0
Venezuela	18.6	56.6	12.3	31.2	2.3
Germany, Fed. Rep.	31.8	28.2	48.0	23.8	15.3
Japan	34.1	24.7	47.1	28.2	16.1
United States	22.5	23.5	45.3	31.2	10.2

Source: Own calculations. For data sources see Appendix C.

Table 2.3: Structural indicators of the capital goods sector, 1985-1987

Country	Range of activities[1] (1)	Metal products[2] (ISIC 381) (2)	Non-electrical machinery[2] (ISIC 382) (3)	Electrical machinery[2] (ISIC 383) (4)	Transport equipment[2] (ISIC 384) (5)	Professional instruments[2] (ISIC 385) (6)
Argentina	20	.27	.19	.16	.36	.02
Brazil	92	.17	.30	.25	.27	.01
Chile	26	.45	.16	.22	.16	.01
Columbia	67	.29	.14	.26	.26	.05
Ecuador	n.a.	.43	.03	.36	.13	.05
Indonesia	38	.29	.07	.23	.41	.00
Malaysia	9	.13	.08	.60	.17	.03
Mexico	31	.22	.19	.20	.33	.05
Peru	39	.26	.17	.26	.27	.03
Philippines	38	.17	.11	.58	.12	.03
Uruguay	n.a.	.26	.07	.24	.41	.01
Venezuela	17	.31	.13	.21	.33	.02
Germany, Fed. Rep.	114	.13	.31	.26	.27	.03
Japan	109	.14	.27	.33	.22	.04
United States	109	.14	.27	.24	.27	.09

Source: Own calculations and UNIDO(1988b). See Appendix C.

1 The range of activities in the capital goods sector refers to the year 1984 and is reported by UNIDO (1988b, pp.126-127). The maximum range of activities is 145.

2 Share in manufacturing value added of capital goods sector (sum of the five industries' outputs), constant 1980 prices. Rounding errors account for the fact that the sum of the shares is not always equal to 1.

products industry are relatively low. In the three industrialized countries included in Table 2.3, activity in this industry accounts for only a small share of total MVA of the capital goods sector. The finding is in sharp contrast to the importance of metal products fabrication in most developing economies, where it contributes more than 25% to value added of all capital goods industries in eight of the twelve countries looked at. The capital goods sector of Chile and Ecuador are even dominated by this low-technology intensive industry.

The industry requiring the most advanced technological skills within the capital goods sector is the non-electrical machinery industry (ISIC 382), in which engines and turbines, metalwork, woodwork, and agricultural machinery are manufactured. The entry into production of these complex capital goods is difficult since it requires skills in design, product engineering, and manufacture. Developing countries which have started production of complex capital goods have relied to a significant extent on transfer of technology agreements in order to obtain design and manufacturing technology (Chudnovsky et al. 1983, p.130). Besides technological barriers, the impact of the market size is also a determining factor for domestic production, as the range of products in the non-electrical machinery industry is large and production does not occur according to anticipated demand but to orders. An inspection of the third column of Table 2.3 shows that the manufacture of these high-technology and skill-intensive products acquires an important role in the industrialized countries, where they account for at least a fourth of value added of the capital goods sector. Their role in the developing countries is strikingly different. Only in Brazil does value added in the capital goods sector originate to more than 20% in the non-electrical machinery producing industry.

Activity in the electrical machinery industry (ISIC 383) centers around the production of generators, transmitters, lamps, and electrical fixtures, but also consumer durables such as radios and television sets. Of all the manufacturing industries, it is both the one growing fastest in the world (average of 6.5 per cent growth from 1975 to 1985) and in developing economies (7.9 per cent in the period from 1975 to 1985).[14] Once the early phase of crude copying and adaptive imitation of previously imported goods (which are generally made in the metal products and transport equipment industries) has passed, the electrical machinery industry offers considerable growth potential, especially in the manufacture of radios and televisions (often for export). Production of heavy electrical equipment like generators, switchgear, and transformers is usually taken up last within this industry. As is shown in Table 2.3, in both Malaysia and the Philippines the industry contributes more than half of total value added accruing in the capital goods sector. The structure in both countries reflects the important role of exports of electrical equipment.

---

14 See UNIDO (1988b, p.108).

A number of developing countries attempt a policy-induced entry into capital goods production by stimulating activity in the transport equipment industry (ISIC 384; shipbuilding and repair, railroad equipment, motor vehicles, motor cycles, and bicycles). For example, vehicle assembly in completely knocked-down kits (all parts imported) is used as an entry opportunity.[15] The technological requirements of assembly are relatively low, since the assembly itself does not require technological skills to a great extent. Only a few countries supply parts and components in the manufacture of transport equipment locally, namely Brazil, the Republic of Korea, and Mexico (UNCTAD 1985a, p.68), since this requires an established iron and steel, metal products, and non-electrical machinery industry. It is probably the industry in which least can be said about the skills involved and its role in establishing a capital goods sector. Some countries use it to develop indigenous techniques at early stages in the development of capital goods industries; for others, it has already acquired an important long term role in the export structure.

Finally, the professional and scientific equipment industry (ISIC 385) is a high-technology industry, but essential for mechanical engineering producers, providing measuring and control instrumentation. Its contribution to value added in the capital goods sector is relatively small both in the developing as well as in the three industrialized countries examined here.

The 'range of activities' in the capital goods sector as displayed in the first column of Table 2.3 refers to a grouping of all different capital goods according to a relative homogeneity of the technological requirements in the production processes.[16] As can be observed, it shows a considerable degree of variation. While all three developed market economies have the broadest range (more than 100 technologically different processes), some developing economies show surprisingly few activities, especially Malaysia. It emerges here that a relatively strong capital goods sector (5.4% of Malaysian GDP originated in the capital goods industries in 1985/87) is not necessarily associated with a broad range of activities. Malaysia's export-oriented development strategy has placed considerable importance on the capital goods sector in general and the electronics industry in particular, which accounts for about a third of total export earnings (UNIDO

---

15 The UNIDO (1988b, pp.126-127) survey reports activity in the assembly of passenger vehicles, coaches, and buses in 38 developing countries in 1984. Twelve of these countries produce less than ten other capital goods, another eleven between 10 and 30 capital goods. The majority of developing countries engaged in the assembly of buses and passenger cars exhibit a narrow range of other capital goods activities and have chosen this form of entry in order to get acquainted with production techniques. In most cases production is conducted with the participation of transnational corporations.

16 See UNIDO (1988b, pp.122-128) for an exact description of how the technological requirements are determined.

1987, p.204). This specialization in internationally competitive industries also implies that the country does not strive for self-sufficiency at any cost but rather attempts to benefit from the increasing international division of labor within the capital goods sector.

Besides the technological requirements and development strategy, the size of the internal market is also an important determinant of the broadness of domestic production. Indonesia, for example, faces a demand for a wide variety of capital goods associated with its large domestic market which translates into a diverse production pattern. Although the share of GDP stemming from the capital goods industries is shown to be only 1.7% in the 1985/87 period, this nevertheless translates into real capital goods production which is about eight times as high as that of Ecuador and four times as high as that of Chile.

2.4.3. Imports and domestic production of capital goods

After having looked at some of the features of capital goods production and structural differences within the manufacturing and the capital goods sector between the developing and industrialized countries, we now examine the contribution of imported goods to the total supply of the broad product category.

For this purpose, we match trade and industrial statistics (see Appendix C, Table C.1). Since only the broad category of capital goods is of interest here, matching takes place on a relatively aggregate level. Additionally, it should be pointed out that the manufacture of some durable consumer goods also occurs in the industries treated here as producing predominantly capital goods. The definition of SITC entries that correspond to the machinery and equipment producing sector therefore differs from the precise one that represents capital goods imports only (compare Appendix C, Tables C.1 and C.2). In order to analyze production and trade for all countries, only trade flows between the developed market economies (DMECs) and the respective developing country are taken into account since the DMECs report merchandise goods transactions with all partner countries on a continuous basis till 1987. But as South-South trade in capital goods is virtually non-existent, these figures (from UN *Comtrade* Statistics) can be expected to cover trade flows to a high degree. Results of the matching are presented in Table 2.4.

Table 2.4: Imports and domestic production of capital goods, 1976-1987

Country	Apparent consumption[1] (1979/ 1981=100) (1)	Domestic production[2] (1979/ 1981=100) (2)	Real imports[3] (1979/ 1981=100) (3)	Share of imports in apparent consumption (per cent) (4)	Self-sufficiency[4] (per cent) (5)
Argentina					
1976-1978	119.6	130.4	46.7	5.2	95.1
1979-1981	100.0	100.0	100.0	13.3	87.2
1982-1984	88.0	94.8	44.7	6.8	94.0
1985-1987	80.5	86.8	39.4	6.5	94.0
Brazil					
1976-1978	93.5	91.5	102.8	11.5	90.2
1979-1981	100.0	100.0	100.0	10.5	92.1
1982-1984	83.7	86.2	77.4	9.7	94.9
1985-1987	94.5	98.6	86.2	9.6	96.2
Chile					
1976-1978	81.6	112.0	55.1	36.4	64.1
1979-1981	100.0	100.0	100.0	53.9	46.7
1982-1984	58.7	57.5	60.0	55.1	45.7
1985-1987	74.5	84.2	66.8	48.3	52.8
Columbia					
1976-1978	78.6	88.8	65.8	37.8	62.7
1979-1981	100.0	100.0	100.0	45.1	55.5
1982-1984	92.9	88.9	98.0	47.5	53.1
1985-1987	95.9	112.5	75.3	35.3	65.1
Ecuador					
1976-1978	77.9	65.2	83.3	75.2	25.3
1979-1981	100.0	100.0	100.0	70.3	30.2
1982-1984	69.8	97.3	59.5	59.8	42.1
1985-1987	80.4	110.7	67.9	59.4	41.6
Indonesia					
1976-1978	81.8	56.5	93.0	78.4	22.3
1979-1981	100.0	100.0	100.0	69.0	32.2
1982-1984	135.6	134.6	136.7	69.6	32.0
1985-1987	123.2	211.7	81.2	45.4	55.4
Malaysia					
1976-1978	48.6	53.5	44.2	61.0	65.2
1979-1981	100.0	100.0	100.0	67.1	59.2
1982-1984	117.6	116.7	137.1	78.2	58.7
1985-1987	92.6	117.3	102.8	74.2	74.7

Table 2.4 (cont.)

Country	Apparent consumption[1] (1979/ 1981=100) (1)	Domestic production[2] (1979/ 1981=100) (2)	Real imports[3] (1979/ 1981=100) (3)	Share of imports in apparent consumption (per cent) (4)	Self-sufficiency[4] (per cent) (5)
Mexico					
1976-1978	63.4	70.9	50.8	30.1	79.8
1979-1981	100.0	100.0	100.0	37.6	71.4
1982-1984	72.3	83.9	74.8	38.9	82.9
1985-1986[5]	68.8	86.1	86.7	47.4	89.3
Peru					
1976-1978	82.0	92.8	66.7	34.4	66.0
1979-1981	100.0	100.0	100.0	42.2	58.3
1982-1984	77.5	71.6	86.1	46.9	53.9
1985-1987	63.7	67.9	58.7	38.8	62.2
Philippines					
1976-1978	83.3	80.3	74.4	59.8	47.0
1979-1981	100.0	100.0	100.0	67.0	48.7
1982-1984	78.7	84.8	97.8	83.3	52.5
1985-1987	45.5	57.3	61.3	90.3	61.4
Uruguay					
1976-1978	74.5	81.5	58.1	24.4	76.0
1979-1981	100.0	100.0	100.0	31.3	69.5
1982-1984	43.6	42.0	46.9	33.7	67.0
1985-1987	57.6	64.0	44.1	23.9	76.9
Venezuela					
1976-1978	106.1	95.6	117.2	53.4	46.9
1979-1981	100.0	100.0	100.0	48.3	52.0
1982-1984	79.9	83.6	76.0	45.9	54.4
1985-1987	78.5	91.2	65.6	40.3	60.4

Source: Own calculations. For data sources see Appendix C.

1 Apparent consumption is defined as production minus exports plus imports.

2 Domestic production of capital goods (ISIC 381, 382, 383, 384, 385), constant 1980 dollars.

3 Imports of capital goods, constant 1980 dollars (see Appendix C, table C.1 for a definiton in terms of the Standard International Trade Classification). Only the trade flows between the developing economy and the developed market economy countries are taken into account.

4 Domestic production over apparent consumption.

5 Industrial statistics for Mexico are incomplete for 1987.

The first column shows the apparent consumption of capital goods, which is defined as domestic production minus exports plus imports.[17] The figure is also termed 'hardware installation', since it measures machinery and equipment that is installed in the economy.[18] The second and third columns contain indices of domestic production and real imports of capital goods. The importance attached to imports in the total supply of machinery and equipment is shown in the fourth column, which sets them in relation to apparent consumption. One might expect that countries with a strong manufacturing base and a high share of capital goods in total industrial output also show a low contribution of imports to total consumption, but this need not be the case. A country following an export-oriented development strategy might import a substantial part of installed machinery and equipment in exchange for the capital goods in which it has specialized in and which it exports. Two countries with the same structural characteristics can thus exhibit different values for the share of imports in apparent consumption. This ratio can be contrasted to the self-sufficiency ratio (column five), which sets domestic production in relation to domestic consumption. A country which exports a substantial portion of its domestic production, but imports about the same value of machines appears to be 'self-sufficient' although a large share of installed machinery stems from abroad. For a country with little or no exports compared to domestic production, the two ratios add up to about 100. An inspection of these two columns again stresses the important role of exports in the Malaysian, but also the Mexican and Philippine capital goods sector.

Imports of capital goods played a major role in installed machinery for all countries except for Argentina and Brazil, where they accounted for less that 10% in 1985/87. In fact, close to half or more of apparent consumption of capital goods originated in the industrialized countries in Chile, Ecuador, Indonesia, Malaysia, the Philippines, and Venezuela.

As can be observed from the table, the sharp import declines in the 1982/84 period went hand in hand with a drop in domestic production in all countries. No country was able to raise domestic production of investment goods, which would have at least partly compensated for the decline in the foreign supply of capital goods. The general feature that both production and imports moved in line with each other continued for most of the countries in 1985/87, when imports and domestic output recovered in some economies (Brazil, Chile, Ecuador, Mexico), while they

---

17 Inventory changes are not taken into account.

18 The index of hardware installation can only serve as a proxy since, as mentioned above, the industries classified as producing investment goods also manufacture some consumer durables. The proxy for hardware installation (derived from production statistics) differs from gross domestic investment of the economy (derived from national account statistics) conceptually. For example, infrastructural investment or construction outlays are only taken into account in the 'gross domestic investment' measure.

both dropped further in others (Argentina, Peru, the Philippines). Three cases can be set forth against this homogeneous change in imports and domestic supply: Columbia, Indonesia, and Malaysia report higher domestic production of capital goods while imports fell off steeply in 1985/87. The same can be observed, although to a somewhat smaller extent, for Uruguay and Venezuela (1985/87).

The composition of capital goods installation underwent marked alterations. In the presence of foreign exchange shortages, the import content of new investment was cut back in most economies. A shift in the composition of investment in favor of domestically originating capital goods can be due to substitution efforts in specific projects, but it is also possible that those investments were implemented which required a small foreign capital content, thereby lowering the overall contribution of imported investment goods to hardware installations. Mexico and the Philippines, on the other hand, show a rising share of capital formation consisting of machines manufactured in the industrialized countries at times when both production and real imports were decreasing.

2.4.4. An assessment

A mere comparison of trade and industrial data does not allow us to judge whether imported capital goods were 'substituted' for by domestic ones. Substitution is a function not only of practicability, but also of need. Even if possible, it does not occur if relative price changes do not induce its realization. But the nature of capital goods production in combination with the empirical findings we made above suggests factors which impede a *perfect* substitutability:

- We observed that production structures in developing and developed countries differ significantly from each other (Tables 2.2 and 2.3). The divergence is most pronounced in the relatively low technology-intensive metal products industry which acquires a much higher share in value added of the capital goods sector in the developing than in the industrialized countries. Furthermore, while activity in the high-technology non-electrical machinery industry is significant in the latter, it accounts only for a minor share in most developing countries studied here; demand for these products is largely met by imports. Capital goods produced in industrialized countries differ from those manufactured in LDCs due to technological requirements which can act as a barrier of entry to the manufacture of previously imported goods.[19]

---

19 Ikemoto (1986) studies the sources of economic growth in Asian developing countries from 1970 to 1980; he examines the contribution of labor, domestic capital input, imported capital input, and time. He concludes that imported capital inputs have a positve effect on total output growth independent of domestic capital inputs. See also Chudnovsky et al.

Adjustment costs associated with the acquisition of technological knowledge are high (and adjustment time long) as it involves training and relocation of the labor force, revaluation of investment plans, reconstruction of machines already in place, and the design and manufacture of new capital goods. The majority of countries studied here rely heavily on imported capital goods in total capital formation and produce only a limited range of products compared to the industrialized countries (Tables 2.3 and 2.4). Due to the very nature of capital goods production, the development of local machinery and equipment producing facilities is complicated:

"*...the entry* [into capital goods production] *is not a one-time affair - of building a plant or plants - but rather a continuing process. It is a process in which a whole structure of complementary industries needs to be built in order to permit the achievement of the economies of specialization as well as those of scale which are characteristic of the machine-building activities*" (Chudnovsky et al. 1983, p.78).

2.5. Intermediate goods

It is convenient to shed some light on the importance of imported intermediate goods and their domestic counterparts by distinguishing primary raw materials from industrial intermediate goods.

2.5.1. Primary intermediate goods

The examples cited above when stressing the role of non-traded specific factors (land sites, climate conditions) for imperfect substitutability for imported goods apply largely to imported raw materials such as unmilled wheat and rice, natural rubber, raw cotton, metalliferous ore, coal, crude petroleum, or natural gas, which are mainly used as inputs in the food-processing and heavy industrial supplies producing industries. Van Bochove (1982, pp.55-57) presents a rough estimate of the share of imports of primary products that is caused by lack of specific factors of production. All primary inputs are included in his calculation for which domestic production falls short of representing at least 20% of total imports of the corresponding commodity. As he finds the

---

(1983), Stewart (1977), and UNCTAD (1981, 1985a, 1988a). UNCTAD (1988a, p.17) concludes that the period since the outbreak of the debt crisis is too short to have permitted local firms to replace foreign technology suppliers on a significant scale. In addition to the direct impact stemming from the different embodied technology, Katrak (1985) finds that imports of machinery and equipment are positively related to expenditures of firms on research and development in developing countries.

Table 2.5: Intensity of manufacturing production with respect to imported industrial raw materials

Country	Share of imported industrial raw materials in manufacturing output[1] (per cent)	Share of imported industrial raw materials in total intermediate imports (per cent)
Argentina		
1979-1981	1.1	24.1
1985-1987	0.6	23.7
Brazil[2]		
1979-1981	7.5	62.0
1985-1987	4.9	62.2
Chile		
1979-1981	10.1	47.4
1985-1986	4.1	36.9
Columbia		
1979-1981	2.8	16.2
1985-1987	1.8	13.4
Ecuador		
1979-1981	2.8	9.2
1982-1984	3.2	9.6
Indonesia		
1979-1981	9.6	18.1
1985-1986	7.8	24.6
Malaysia		
1979-1981	10.6	23.2
1985-1987	7.2	15.7
Mexico		
1979-1981	2.3	20.3
1985-1986	2.0	24.2
Peru		
1979-1981	2.4	21.8
1982-1984	3.0	20.5
Philippines[2]		
1979-1981	14.5	44.0
1985-1986	12.7	42.6
Uruguay[2]		
1979-1981	7.7	47.2
1985-1987	6.5	46.5
Venezuela		
1979-1981	2.4	13.3
1985-1986	2.6	18.1

Source: Own calculations. For data sources and method of aggregation and deflation see Appendix C.

1 Import and domestic manufacturing output figures in 1980 constant dollars.

2 Countries that were net importers of industrial raw materials in 1985/1987 (for the Philippines 1985/86).

vast majority of total primary imports to fall under this criterion, he concludes: "If primary products are imported, this happens because the country lacks specific factors necessary to produce them domestically" (Van Bochove 1982, pp.56-57). An indicator of the role these essential imports of primary raw materials acquire in the twelve developing countries under study here is their use as inputs in the manufacturing process. Table 2.5 shows the intensity of use of manufacturing output with respect to imported raw materials and the share the latter hold in total imports of intermediate goods.[20]

It emerges that the input-output ratios for imported primary raw materials in manufacturing output are highest for the Philippines, Malaysia, Indonesia, and Chile. While the intensity of use of imported primaries declined considerably in some countries in the 1980s (Brazil, Chile, and Malaysia), it remained relatively constant for the majority of them. The importance of primary raw materials is also manifested by their weight in total intermediate imports. Close to two thirds of imported intermediates consisted of primary raw materials in 1985/87 for Brazil, somewhat less than half for Uruguay and the Philippines, and about a third for Chile. In these economies, non-traded specific factors claim responsibility as the driving force for a substantial part of total imports.

2.5.2. Industrial intermediate goods

2.5.2.1. Nature of production

Some industries producing industrial intermediate goods can be characterized as 'light' industries, such as the wood and wood products industry (ISIC 331) or the rubber products industry (ISIC 355), but the bulk of activity occurs in 'heavy' industries manufacturing e.g. industrial chemicals (ISIC 351), petroleum products (ISIC 353), iron and steel (ISIC 371), or non-ferrous metals (ISIC 372).[21]

To a large extent the intermediate goods' industries are process industries meeting a broad demand from upstream industries and can produce according to anticipated demand rather than to custom orders as is the case in most manufacturing processes in the capital goods sector. Examples are to be found especially in the heavy industrial supplies production (iron and steel,

---

20 Compare Appendix C, Table C.3 for a definition of primary raw materials in terms of the Standard International Trade Classification.

21 See Kirkpatrick et al. (1984, pp.16-18) for a distinction between and description of 'light' and 'heavy' industries. See Table C.1 in Appendix C for a complete list of industries producing predominantly industrial intermediate goods.

industrial chemicals). These industries are characterized by high forward linkages to the industries producing consumer goods as well as to the capital goods sector. At the same time they make intensive use of raw materials. For example, the paper and paper products industry, which manufactures pulp, paper, containers, and boxes, is heavily dependent on raw wood and basic chemicals but also has strong forward linkages to the food processing plants. Similarly, the industrial chemicals industry (fertilizers, pesticides, plastic, synthetic fibres, rubber, and organic chemicals) is strongly linked to inputs of crude oil, phosphate, wood, cotton fibres, and the like. Strong forward linkages can be observed to the consumer goods producing and agricultural sector. The sensitivity of these industries to foreign exchange availability depends on the extent raw materials have to be imported. For example, the reliance of the iron and steel industry (producing rolled steel sheets, castings, forgings, and pipes) on imported inputs is a function of domestic depletable iron and ore deposits.

Technology requirements vary widely for industrial intermediates manufacturing operations. Among the thirteen industrial intermediates industries, eight can be classified as low-technology intensive, while four operate in a medium-technology environment. But the industrial chemicals industry as well as some branches of the plastic industry, namely drugs (ISIC 3522) and the plants manufacturing telecommunications equipment (ISIC 3560), are characterized by high-technology intensive production processes.[22]

2.5.2.2. Imports and domestic production of industrial intermediate goods

Table 2.6, structured in the same way as Table 2.4 for capital goods, gives results if we match import and domestic production data.[23] The self-sufficiency ratio (column 5) indicates that some of the countries are net exporters of industrial intermediates. Malaysia, Indonesia, and Venezuela specialized in exports of refined oil products (and tin products in the case of Malaysia), while importing others so that foreign supplies of industrial intermediates account for a substantial portion in apparent consumption. Overall, industrial intermediate goods account for a smaller portion in supply than imported capital goods. Ecuador, Indonesia, Malaysia, the Philippines, and Venezuela report the highest shares of imports in the apparent consumption of industrial intermediate goods.

---

22 Compare UNCTAD (1987a).

23 Compare Appendix C, Table C.1.

Table 2.6: Imports and domestic production of industrial intermediate goods, 1976-1987

Country	Apparent consumption[1] (1979/ 1981 = 100) (1)	Domestic production[2] (1979/ 1981 = 100) (2)	Real imports[3] (1979/ 1981 = 100) (3)	Share of imports in apparent consumption (per cent) (4)	Self-sufficiency[4] (per cent) (5)
Argentina					
1976-1978	120.3	122.0	64.0	2.2	98.5
1979-1981	100.0	100.0	100.0	4.2	97.1
1982-1984	124.4	127.2	63.4	2.1	99.4
1985-1987	120.5	123.6	55.9	1.9	99.6
Brazil					
1976-1978	82.7	80.7	112.2	5.2	96.6
1979-1981	100.0	100.0	100.0	3.8	98.9
1982-1984	96.9	101.2	56.3	2.2	103.2
1985-1987	116.7	122.0	66.9	2.2	103.4
Chile					
1976-1978	75.6	79.3	56.2	8.0	121.3
1979-1981	100.0	100.0	100.0	10.8	115.7
1982-1984	101.5	107.1	80.2	8.5	122.0
1985-1987	166.5	161.8	87.0	5.6	112.4
Columbia					
1976-1978	88.2	89.3	81.4	12.2	92.6
1979-1981	100.0	100.0	100.0	13.2	91.4
1982-1984	98.1	98.1	104.8	14.1	91.4
1985-1987	122.2	126.5	108.4	11.7	94.6
Ecuador					
1976-1978	82.2	76.7	86.9	33.0	74.2
1979-1981	100.0	100.0	100.0	31.2	79.6
1982-1984	116.8	113.0	113.3	30.3	77.0
1985-1987	127.3	135.3	96.9	23.8	84.6
Indonesia					
1976-1978	57.2	58.0	73.0	45.4	98.9
1979-1981	100.0	100.0	100.0	35.6	97.5
1982-1984	125.2	122.7	115.1	32.7	95.5
1985-1987	184.3	215.9	91.8	17.7	114.2
Malaysia					
1976-1978	51.6	64.0	48.4	20.9	134.5
1979-1981	100.0	100.0	100.0	22.2	108.3
1982-1984	120.0	109.3	121.7	22.6	98.7
1985-1987	124.2	121.8	110.9	19.9	106.3

Table 2.6 (cont.)

Country	Apparent consumption[1] (1979/ 1981=100) (1)	Domestic production[2] (1979/ 1981=100) (2)	Real imports[3] (1979/ 1981=100) (3)	Share of imports in apparent consumption (per cent) (4)	Self-sufficiency[4] (per cent) (5)
Mexico					
1976-1978	78.1	81.1	53.6	8.1	96.3
1979-1981	100.0	100.0	100.0	11.7	92.7
1982-1984	99.8	106.7	77.3	9.1	99.1
1985-1987	121.6	131.6	83.3	8.0	100.3
Peru					
1976-1978	81.7	80.0	75.4	10.6	107.8
1979-1981	100.0	100.0	100.0	11.5	110.1
1982-1984	74.5	82.0	81.0	12.5	121.3
1985-1987	92.0	100.5	85.1	10.7	120.1
Philippines					
1976-1978	87.4	84.8	96.2	17.7	88.9
1979-1981	100.0	100.0	100.0	16.1	91.6
1982-1984	91.8	93.7	91.6	16.1	93.5
1985-1987	74.1	78.3	75.1	16.3	96.8
Uruguay					
1976-1978	100.8	102.4	74.6	5.4	97.9
1979-1981	100.0	100.0	100.0	7.3	96.4
1982-1984	78.4	80.3	63.2	5.8	98.7
1985-1987	91.2	92.5	80.2	6.4	97.8
Venezuela					
1976-1978	86.2	90.1	101.1	26.4	134.4
1979-1981	100.0	100.0	100.0	22.5	128.6
1982-1984	77.5	94.1	89.1	25.9	156.3
1985-1987	84.1	109.7	73.9	19.8	167.8

Source: Own calculations. For data sources see Appendix C.

1 Apparent consumption is defined as production minus exports plus imports.

2 Domestic production of industrial intermediate goods (ISIC 321, 331, 341, 351, 352, 353, 354, 355, 356, 362, 369, 371, 372), constant 1980 dollars.

3 Imports of intermediate goods, constant 1980 dollars (see Appendix C, Table C.1 for a definiton in terms of the Standard International Trade Classification). Only the trade flows between the developing economy and the developed market economy countries are taken into account.

4 Domestic production over apparent consumption.

If one compares production and trade data (columns two and three), it emerges that most countries raised domestic output in times imports were curtailed in the 80s; the import cut was fully offset in Argentina (82/84), Chile (82/84), Ecuador (85/87), Indonesia (85/87), Malaysia (85/87), and Venezuela (85/87). Consequently, the share of imports in domestic use declined in most countries, most drastically in Indonesia. More than a third of the country's supply of industrial intermediates originated abroad in 1979/81, but as production increased twofold till 1985/87 and imports dropped, the ratio declined to 17.7%.

2.5.3. An assessment

As was the case for capital goods, we cannot conclude from trade and production data whether a 'substitution' took place or not in the developing economies in the 1980s. But shortly summarizing the description of primary and industrial intermediate goods, we can ascertain that an imperfect substitutability of the bundle of imported intermediate products is caused by

- non-traded factors of production. They motivate imports of primary raw materials, which account for a substantial portion of total trade in intermediate goods for some economies. We observed that imported industrial raw materials are an important input into manufacturing production for several countries (Table 2.5).
- technological requirements associated largely with production processes in the chemical and plastic products industries;
- adjustment costs and time. Although we found that self-sufficiency ratios are much higher in industrial intermediate goods than in capital goods, several of the countries are strongly engaged in intra-industry trade (Table 2.6). Adjustment costs and time are obstacles to be overcome particularly in starting production in the process industries (iron and steel, chemicals, rubber, and non-ferrous metals).

2.6. Development strategies and the role of imports

As noted above, the role of imports in the total supply of capital and industrial intermediate goods varies among the developing countries studied here; the dispersion is pronounced in the case of capital goods. This can partly reflect the size of the economies, natural resource endowments, and the lack of specific factors of production, but a strong influence is also exerted by the development strategy adopted. This section briefly summarizes the main thrusts of an inward- and an outward-oriented development strategy.

2.6.1. Inward- versus outward-oriented policies

The Great Depression and the Second World War left a lot of developing countries pessimistic with regard to the role of trade in economic development. As commodity export prices and quantities were strongly influenced by the world wide depression, primary commodity exports came to be regarded as not living up to a role of being an 'engine of growth'. Additionally, the supply of manufactured goods from the industrialized countries decreased substantially during the Second World War. Developing countries consequently began to strive for increased self-sufficiency in the supply of manufactured goods, which implied industrialization. South European and Latin American countries started such a drive in the 1930s and 1940s, with most Asian developing countries joining in during the 1950s. With the exception of a few countries (namely Malaysia, Singapur, and Hong Kong), this industrialization was to be achieved through deliberate import substitution policies, aimed at fostering the manufacturing sector in developing countries by providing emerging industries with a protective shield against foreign competitors.

By deliberate trade policies the domestic production of previously imported goods was to be taken beyond the 'natural' rate of import substitution that every growing economy experiences as human skills are acquired and the comparative advantage in international trade changes. The industrialization strategy which was followed by most countries came to be known as the 'theory of development stages': "....most advocates conceded that import substitution should begin with the domestic production of consumer goods. Over the long term, the approach was pictured as a sequential process working its way from light consumer goods into the production of intermediate products and industrial supplies and culminating in the domestic production of capital goods" (Ballance and Sinclair 1983, p.37). Domestic production of non-durable consumer goods exploited a ready-made market created by imports and it was thought that the establishment of these industries would automatically create backward linkages through their demand for intermediate and capital goods. Consequently, most countries levied high nominal tariffs and quantitative restrictions, rising with the stage of manufacture so that effective protection rates were even higher than the nominal rates. Consumer goods were targeted with the highest protective devices or banned altogether, while imports of capital goods faced a much lower rate of protection and were often even exempted by means of draw-back provisions and fostered by an overvalued exchange rate. Interest rate ceilings were established, going hand in hand with credit rationing in order to channel preferential credit to 'essential industries'. Other instruments ranged from granting tax incentives, exerting price controls, providing direct subsidies for investment to active industrial investment planning.

These policies automatically created an anti-export bias since production of import substitutes was more profitable than engagement in export production. Following Krueger (1985, p.20) and Sachs (1989, p.14), an 'inward-oriented trade regime' can be characterized as one in which there is less incentive to earn than to save foreign exchange.[24] Or, stated differently, one in which incentives to engage in production of import substitutes for the domestic market are larger than in production for exports. An 'outward-oriented trade regime', on the other hand, can be defined accordingly as being characterized by at least as much incentive to earn as to save foreign exchange. Hence, an outward-orientation can be noted for countries that pursue liberal trade and exchange rate policies in order to not bias the incentive structure against exports, but also for countries that employ trade policy measures in order to protect emerging industries, but that are at the same time careful to at least neutralize the anti-export bias created by import-substitution policies. It is by now well established that some South-East Asian countries, notably the Republic of Korea, have pursued this latter strategy of assigning the state an important role in developing and protecting domestic industries but at the same time subsidizing exports and supporting them by a realistic exchange rate. Similarly, credit rationing was not targeted to support production for the domestic market but to discriminate in favor of exports.[25]

The adoption of outward-looking policies is often characterized by the easing of quantitative restrictions, replacement of quotas by tariffs and a reduction of the dispersion in the protective structure. The latter reduces the distortion in the manufacturing sector, and provides incentives for the production of intermediate and capital goods. Additionally, cost disadvantages of the export sector are neutralized by remission of customs duties on imported materials and components used in the manufacture of exports, income tax exemptions, and direct subsidies.

We summarize briefly the main policies adopted by the twelve developing countries in Table 2.7. As can be noted, turning-points in the trade regime occur for most countries,[26] which suggests that a clear-cut distinction between 'outward-orientation' and 'inward-orientation' does not exist and that trade policies should be rather viewed as lying on a continuum of possible regi-

---

24 Compare Bhagwati (1978), Donges (1976), Little et al. (1970), and Krueger (1978) for the initial discussion of inward- and outward-orientation.

25 For the analysis of the Korean development strategy compare, e.g., Collins (1988), Collins and Park (1989), Dornbusch and Park (1987), Sachs (1987), and UNIDO (1983, pp.126-130).

26 Donges (1976) and Donges and Riedel (1977) note turning points in the trade policy regime for most developing countries they examine in the late 1960s or beginning of the 1970s. A marked shift in trade orientation towards a more outward-oriented development in the mid-1980s is emphasized by Laird and Nogues (1988a, 1988b), UNCTAD (1989, Chapter 5), Whalley (1989), and Williamson (1990).

Table 2.7: Trade and industrial policy orientation

Country	Trade and industrial policy orientation
Argentina	Development strategy which fostered production for the domestic market was pursued since 1930s. Move into secondary import substitution for heavy intermediate and capital goods occurred in the late 1960s, when support for export industries showed little success as the general inward-oriented strategy was not abandoned. The liberalization attempt from 1976 to 1981 was eroded by high inflation rates and postponed real exchange rate adjustments. Highly protective trade policies were introduced in response to the severe balance of payments crisis in 1981. Reliberalization and export promotion started in 1987, including e.g. improved credit availability and lending conditions for export financing.
Brazil	The traditional import-substitution strategy to support industrialization and supply the domestic market was reformulated in 1962, when a more open development policy was adopted. A reversal to secondary import substitution took place in 1974, especially targeted at supporting domestic machinery and equipment manufacturers; the protectionist pattern was altered in favor of supporting capital goods production. Protectionist policies were tightened during the debt crisis. A rationalization of the tariff structure and a reduction of tariff rates occurred in 1988, but import licenses remained binding for many products.
Chile	A strong inward-oriented development policy was followed till 1974, when a major policy reversal based on liberal trade and industrial policies occurred. High inflation rates in the presence of an exchange rate pegged to the dollar led to an overvalued exchange rate and high current account deficits at the end of the 1970s and beginning of the 1980s. Tariffs were increased in response to the debt crisis, but lowered hand in hand with adjustment policies in 1985.
Columbia	Change of trade policy from inward-orientation to export promotion took place in 1967, including modest trade liberalization, development of a uniform structure of export incentives and outright promotion of non-traditional exports (textiles). While increased protection was relied on to curb import demand at the beginning of the 80s, a stabilization program was enacted in 1984, which included major devaluations, a rationalization of the tariff structure, and a trade policy reform.
Ecuador	The economy benefitted from high oil prices in the 1970s and a vigorous industrialization program (Industrial Promotion Law) was enacted. An overvaluation of the sucre occurred (1972-1982) under the fixed exchange rate regime . Lower oil prices and high interest rates led to the adoption of a conventional stabilization program in 1982, but import controls were also tightened in response to severe balance of payments difficulties. Trade policy and financial market liberalizations started in 1984, but were partly reversed in 1986 in response to the oil-price decline.
Indonesia	A strong liberalization was carried through in 1966, but rates of protection intensified in the 1970s. Balance of payments problems were met by three large devaluations (1978, 1983, 1986) and small depreciations in between, which gave boost to non-oil exports. The large devaluations were accompanied by major tariff reductions, elimination of import quotas, austerity budget policies, an overhaul and reform of the tax system, and a renunciation of the requirement of production licenses.
Malaysia	An outward-oriented development strategy was adopted when independence was achieved, with financial and trade policies supporting exports. Industries which were designed to strengthen non-traditional exports (food, wood products, textiles; later machinery and transport equipment) were granted strong initial protection, but the general level of tariffs remained relatively low and revenue-oriented throughout the 1970s and 1980s. Exchange rate adjustments were relied on to confront the terms of trade deterioration in the mid-80s, while the trade regime was not altered significantly.

Table 2.7 (cont.)

Country	Trade and Industrial Policy Orientation
Mexico	While a promotion of exports started in 1971, the emphasis of industrial and trade policy continued to lie on import-substitution industrialization first targeted towards consumer industries and later extended to capital goods manufacturing. The oil boom at the end of the 1970s encouraged import-liberalization policies, which were short-lived as high borrowings triggered the debt crisis in 1982. Massive devaluations and use of quantitative import controls were enacted from 1982 to 1985. A strong policy reversal can be noted in 1985, targeted predominantly at fostering exports (elimination of prior export approval, easing of export finance).
Peru	The policies in the 1960s and 1970s were targeted at establishing industries supplying the domestic market. A multiple exchange rate combined with high levels of protection against 'inessential' imports and widespread licensing of imports was employed without neutralizing the imposed anti-export bias. A sharp devaluation occurred in 1978, which was combined with strengthening incentives to export and the elimination of a range of import restrictions. The Belaunde government set out to continue these policies, but import prohibitions and a policy reversal to inward-orientation succeeded the initial reform moves.
Philippines	The thrust of postwar Philippine development lay on import substitution, while exports were discriminated against. A major break in trade policy occurred in 1981, when an import decontrol program was started and the Tariff Act was passed. These outward-oriented policies were soon succeeded by the imposition of import controls in response to the balance of payments difficulties encountered in 1983. The spread of protection was reduced in 1985 when the original Tariff Act came into effect.
Uruguay	A widespread liberalization and export promotion took place in the 1970s (reduction of protection, decontrol of prices, encouragement of manufactured exports). The maximum tariff rate and tariff dispersions were reduced in 1975, import quotas and export taxes eliminated. A pre-announced crawling peg used to curb domestic inflation rates created problems in the late 70s. A stabilization program under IMF supervision was carried out in 1983, and a tax reform was enacted together with a liberalization of financial markets in 1985.
Venezuela	The 1970s were marked by high protection of the emerging manufacturing industries, hindering non-oil export expansion. Import restrictions were further fostered in 1983 and the allocation of foreign exchange was required for importation at the official exchange rate. A liberalization started in 1985 when non-traditional exports were shifted to the free market exchange rate and large depreciations in this market provided considerable incentives to export.

Source: For Argentina see Kürzinger (1988), UNIDO (1983, pp.143-145), UNIDO (1987, pp.43-44), and Williamson (1990). Brazil: Cardoso and Fishlow (1989b), Williamson (1990), and World Bank (1983). Chile: Eggers (1988), Williamson (1990), and World Bank (1987a). Columbia: Bautista et al. (1981, pp.99-140), UNIDO (1979, p.89), Williamson (1990), and World Bank (1987b). Ecuador: Bocco (1990) and Williamson (1990). Indonesia: Bhattacharya and Linn (1988), UNIDO (1987, pp.56-58) and Woo and Nasution (1989). Malaysia: Bautista et al. (1981, pp.184-213), Bhattacharya and Linn (1988), Donges (1976), and Hock (1986). Mexico: Buffie (1989), Ebenroth and Jäger (1990), UNIDO (1987, pp.42-43), and Williamson (1990). Peru: Corsepius (1989), Williamson (1990), and World Bank (1985). The Philippines: Bhattacharya and Linn (1988), Dohner and Intal (1989), and Tan (1986). Uruguay: Williamson (1990) and World Bank (1989a). Venezuela: World Bank (1988a).

mes.[27] But if one considers the 1970s and 1980s, the trade and industrial policies of Chile, Columbia, Indonesia, Malaysia, and Uruguay can be broadly classified as more 'outward-oriented', while the development strategy of the remaining countries was rather 'inward-oriented' as the protection of industries was relatively strong and, more importantly, an anti-export bias persisted.[28]

2.6.2. Effects of inward-oriented policies

In most countries pursuing inward-oriented industrialization policies, a typical pattern of industrial development resulted. Countries did become much more self-sufficient in meeting the demand for consumer goods, but in contrast to what was expected at the outset of the industrialization drive, the ratio of imports to total income seldom declined. On the contrary, it often increased since industrialization boosted the demand for various types of industrial supplies and capital goods which were seldom available locally. Additionally, the distorted factor prices (minimum wage legislation plus encouragement to import capital equipment) caused domestic entrepreneurs to prefer imported factors of production to domestic substitutes. As a consequence of the factor-price distortions, protected industries operated with relatively capital-intensive factor combinations as surveys of industries in selected countries revealed.[29] This factor use contradicts what, according to the international trade theory, is thought to be the comparative advantage of the developing countries.

Consequently, the process of inward-oriented import-substitution policies left countries often more vulnerable to foreign exchange shortages than they were at the outset.[30] "As long as an economy imports finished goods, its occasional inability to obtain or pay for these imports lowers the standard of living by making the imports unavailable. When the economy produces the same goods domestically but with the aid of some imported materials, parts, or equipment, an occasional inability to import these will also cause work stoppages, unemployment, and a fall in the income generated by domestic manufacturing" (Little et al. 1970, p.61).

---

27 See Edwards (1989).

28 For a general classification of countries according to their trade and industrial regimes compare e.g. Balassa (1984). See Weiss (1988, pp.35-41) for a critical comment on such classifications. Bhagwati (1988) and Edwards (1989) give comprehensive discussions on trade policies and definitions of development strategies.

29 Compare Bautista et al. (1981), Donges (1976), and Little et al. (1970).

30 For the assessment that an inward-oriented development raises the vulnerability to foreign exchange shortages compare Ballance and Sinclaire (1983, pp.40-41), Cardoso and Fishlow (1989a, p.12), and UNIDO (1983, pp.141-142).

Furthermore, the foreign exchange vulnerability was even more pronounced as the higher reliance on imported intermediate and capital goods was not balanced by a higher degree of export independence. The fostering of manufacturing industries supplying the domestic market created an anti-export bias against traditional primary exports from the agricultural sector and did not lead to an export diversification.

2.6.3. Secondary import substitution

The most serious threat to the strategy was the scope of the domestic market: The 'easy' phase of import substitution involved fostering the industries producing consumer non-durables. At the end of the 1960s, most countries in Latin America and Asia which had launched an inward-oriented industrialization strategy found themselves nearly self-sufficient in the production of these goods and the further growth potential was bounded by the scope of the home market. Further, "with the growth of the modern sector and the rise in the demand for capital goods, it was thought that, unless the policy of import substitution was extended to the capital goods sector, even the policies of import substitution of consumer goods would fail because of an intensified strain on the balance-of-payments" (UNCTAD 1985a, p.146). Some countries consequently moved into secondary import-substitution in order to increase the level of self-sufficiency in the domestic supply of capital and industrial intermediate goods (compare Table 2.7). The policy switch from a relatively outward-oriented development towards inward-oriented secondary substitution policies in 1974 in Brazil was targeted at supporting domestic manufacturers of machinery and equipment. Tariff and non-tariff barriers were increased for imported equipment, which reversed the cascading tariff structure of primary import substitution usually encountered.[31] Fiscal and financial incentives were provided to encourage domestic production of capital goods; the Industrial Development Council supervised investment projects and assured a minimum domestic machinery content in new investment.[32] Argentina and Mexico also started to emphasize domestic manufacture of capital goods.[33]

Malaysia took a different path. Its prime goal in the development of the capital goods sector was not to achieve a high independence from foreign supply, but rather to strengthen it in order to open export opportunities. While initial protection was granted to domestic manufacture

---

31 See World Bank (1983).

32 See Chudnovsky et al. (1983, pp.103-104). See for an example of secondary import substitution Corsepius and Schipke (1989) who examine the protection granted to domestic producers of computers in Brazil.

33 For a description of the Argentinian move into import substitution of capital goods and industrial intermediates see Kürzinger (1988, p.62) and UNIDO (1983, pp.143-145).

of electrical equipment, it successfully avoided the anti-export bias. As mentioned above, exports from the electronics industry now account for about a third of total export earnings. At the same time it did not strive for independence manifesting itself by a curtailment of internationally available capital goods which account for about half of fixed instalment of capital goods, but rather attempted to benefit from the increasing international specialization within the capital goods sector.

We visualize these different development strategies for the capital goods sector in Figure 2.1. The domestic procurement ratio (share of domestically produced capital goods - production minus exports - in apparent consumption) is set in relation to the share of GDP originating in capital goods industries. While most countries, most notably Argentina and Brazil, have reduced the importance of imports in apparent consumption with the strength of the capital goods industries, Malaysia has relied on intra-industry specialization. Its capital goods industries contribute slightly more to overall GDP than in Argentina, but more than half of apparent consumption of capital goods is met by foreign supply.

The diverging structure we observe in Figure 2.1 cannot be explained by factors such as country size or resource endowment alone. This becomes apparent if we add figures for some developed economies. As Chudnovsky et al. (1983, p.3) report, the domestic procurement ratio for a number of small industrialized countries is relatively low, e.g. The Netherlands (55%), Austria (61%), Sweden (60%), and Hungary (70%). But larger countries can also be found to be engaged in considerable intra-industry trade. In West Germany, a net exporter of machinery and equipment where MVA in the capital goods industry was found to contribute more than 15% to GDP in Table 2.2, over one third (34.3.%) of gross fixed capital formation was met by imports in 1985; similarly the United States imports more than a fifth of installed machinery and equipment.[34] These figures if presented graphically in Figure 2.1 would be located above and far to the left of Argentina and Brazil. It can be observed that the self-reliance on capital goods has decreased in the industrialized countries in the past decades, while quite a few developing countries have attempted to increase it.[35]

---

34 Data from UNIDO (1988a).

35 See Jacobsson (1988).

Figure 2.1: Relation between the domestic procurement ratio and the share of GDP originating in the capital goods industries (per cent), 1985-1987[1]

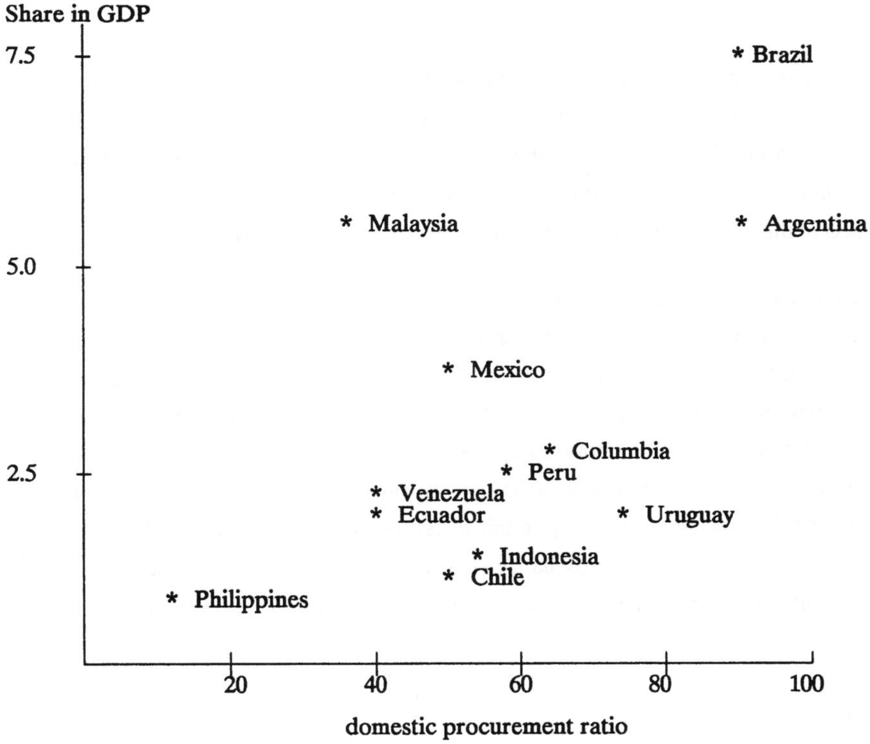

Source: Tables 2.2 and 2.4

1 The domestic procurement ratio is defined as domestic production minus exports over apparent consumption. For Mexico, the domestic procurement ratio relates to the 1985-1986 period only.

2.6.4. Independence or vulnerability?

In the above discussion of development strategies of the capital goods sector, we stressed that some countries implemented secondary import substitution policies in order to reduce the dependence on capital goods manufactured abroad. The self-sufficiency ratio in the supply of investment goods was to be increased. But "self-sufficiency implies a degree of autarchy, of doing without foreign goods, at whatever cost in terms of economic efficiency or the sacrifice of other goals" (Roemer 1981, p.430). Were these costs balanced by a gain in 'independence' as the crucial role of imported capital goods was reduced?

Jacobsson (1988) assigns the substantial and growing intra-industry trade in capital goods to the economies of specialization which can be achieved in capital goods production. He concludes (1988, p.30):

> *"Firstly, the economically efficient limit to import substitution in the capital goods sector has probably been adjusted downwards during the past decades. Secondly, there is a considerable risk that the firms operating within the framework of inward-looking strategies of development of the capital goods sector will in many cases become permanent infants, both with respect to the unit production cost and the technological capability of the firms".*

One crucial matter of investigation of this study is whether this assessment of Jacobsson is supported by empirical evidence. Stated differently, did the developing countries which strove primarily for independence by reducing the share of imported capital goods in new investment achieve a lower vulnerability to foreign exchange shortages by raising the elasticity of substitution between domestically produced and imported capital goods? Or could they not accomplish this aim since the suppressed flow of imported capital goods with its embodied technology remained vital to the economies?[36] And, as a follow-up to this question, did they place themselves in a more vulnerable position in regard to foreign exchange shortages than before, as they often neglected the traditional export sector and discouraged the diversification of the export structure?

2.7. Summary

Three reasons which can impede perfect substitutability for imported goods were put forward in this chapter: non-traded factors of production, lack of technological skills, and adjustment time and costs. We described the nature of production of capital and industrial intermediate goods and compared trade and industrial data. We stressed that besides adjustment costs, which apply to both capital and intermediate goods, the technology-intensity of capital goods production is very high and that non-traded factors of production cause trade in primary intermediate goods. At the end of the chapter, we were concerned with trade and industrialization policies of the twelve developing economies. The question was raised, whether the drive to reduce the importance of imported capital goods in new investment projects achieved its aim of lowering dependency on imported machinery and equipment. The answer to this question has to be postponed until we introduce and estimate production functions.

---

36 The question thus also examines whether the degree of vulnerability to a foreign exchange shortage can be measured by the share of imported products in total supply, as put forward by e.g. UNCTAD (1981, p.11) and UNCTAD (1988a, p.17).

3. Import models: issues and problems

3.1. Introduction

Up to now we have been concerned with institutional and economic factors that impede perfect substitutability between domestic and imported factors of production. The next step is to assess how such an imperfect substitutability can be included in import models, which is the task of this chapter. The discussion of different import models focuses on several questions which are important for our aim of determining the degree of substitutability:

(a) What are the general problems encountered in estimating import demand functions?

(b) Which models take into account imperfect substitutability between domestic and imported production means?

(c) Which models allow us to determine the elasticity of substitution? What are the underlying assumptions of these frameworks and, on that account, which one should be employed in our empirical analysis?

The degree of substitutability for imported factors (and the proper modelling of macroeconomic production functions) is of interest, as theoretical models that explore the influence of imported inputs on the long run equilibrium growth rate of economies derive different results than the standard neoclassical models of trade and growth: The growth rate is found to vary with the degree of substitutability for imported inputs. Results of these theoretical models are briefly summarized in Section 2.[1] The 'standard' import function and modifications thereof are examined in Section 3, while Section 4 contrasts different approaches to model imports as inputs empirically.

3.2. The theoretical background: imports and growth

3.2.1. Perfect substitutability for imports in open economy growth models

The standard neoclassical theory of international trade (Heckscher-Ohlin-Samuelson model) characterizes the international exchange of goods as taking place in final commodities, designated to satisfy consumer demand in the respective importing country. Production functions are assumed to be identical in the trading economies and trade is created through different en-

---

1 Compare Van Bochove (1982, Chapter 2) for a comprehensive survey of growth models that introduce imports as factors of production.

dowments of factors of production (capital and labor).[2] A country gains from international trade by exploiting its comparative advantage - production of the good which employs the relatively abundant factor intensively is increased over consumption and exported. Vice versa, production of the other commodity is reduced and imported. The theory builds on the assumption that imported goods can be perfectly substituted for by domestic ones and that there are no adjustment costs involved in replacing imports by such domestic products.

The characterization of trade along slightly modified lines is first introduced into neoclassical growth theory by Bardhan (1965) and Oniki and Uzawa (1965).[3] These and follow-up studies establish that the basic result from autarchic growth, namely that the long run steady-state growth rate of an economy converges to its natural rate of population growth, is not altered.[4] If imports can be produced domestically and import substitution takes place without adjustment costs, the closed economy result maintains its validity if trade is introduced.

However, this result is altered if the perfect substitutes condition is relaxed and imports are used either as intermediate or final goods in domestic production.[5]

3.2.2. Perfect complementarity in production

Consider a one-good economy characterized by Leontief production technology, which combines the capital stock (K) and imports (M) in fixed proportions to gross output (Q):

$$(3.1) \qquad Q = \min\left[\ k_1K,\ k_2M\ \right],$$

with k_1 and k_2 as the output-input productivities. Domestic saving (S) is linked to total output with

$$(3.2) \qquad S = sQ,$$

2 Trade can be created not only by different factor endowments, but also by differences in labor productivities (Ricardo), total factor productivities, and demand structures. See Siebert (1991, Chapters 3 and 4).

3 Both studies distinguish between trade and production of a consumption and investment good.

4 The restriction has to be introduced that the 'capital-intensity condition' holds - consumption goods must be more capital intensive than investment goods. Compare Bardhan (1965) and Oniki and Uzawa (1965).

5 The description concentrates on the use of imports as factors of production and does not deal with imperfect substitutability in consumption.

while exports are assumed to grow at the exogenously determined constant rate *f*:

$$(3.3) \qquad X = X_0\, e^{ft},$$

with t as a time index. Further, trade is balanced (X=M) and thus investment has to equal savings (I=S). Ignoring depreciations, the change in the capital stock is equal to investment (dK=I). In this small model, the growth rate of the economy (dQ/Q = g) is characterized by

$$(3.4) \qquad g = \min\left[\, f,\ sk_1 \,\right].$$

The autarchic growth rate familiar from the Harrad-Domar growth models (sk_1) can only be achieved if the export growth rate *f* is at least as high as the autarchic growth rate. Further, it need not hold ex ante that imports and exports balance as well as investment and savings, while this has to hold ex post. Either imports or savings will have to adjust downward in order to equalize the two growth rates.[6]

Models of this sort were introduced by Johnson (1953, 1958) and extensions of the framework led to the well-known two-gap planning models for development associated primarily with the name of Hollis B. Chenery and associates (1962, 1966). They postulate a targeted output growth rate for the developing economy and calculate investment and import volumes that are required if the growth target is to be achieved. In combination with the behavioral domestic saving function and export forecasts, a 'savings gap' (I-S) and a 'trade gap' (M-X) are identified and foreign capital in the form of aid or borrowing needed to close the dominant of the two gaps are computed. The important feature of these models is that the domestic rate of growth depends on the outside world, since the latter determines export growth and the availability of foreign funds.[7]

Itsumi and Moriguchi (1978) and Schiavo-Campo (1980) introduce production functions in which capital and labor can be substituted for each other, but show a complementary relation towards imports.[8] Nevertheless, even if such a relaxed assumption is made, a non-saving constraint to growth can exist. Itsumi and Moriguchi show that the long run growth of the economy with ba-

---

6 Compare UNIDO (1984, p.4).

7 For models in the trade gap tradition see McKinnon (1964) and Buffie (1986). This framework has been employed to analyze debt accumulation over time by Farzin (1988), Feder (1981), Hallett (1988), and UNIDO (1984). See Glick and Kharas (1986) and Hentschel (1988) for surveys of these models.

8 Compare also Findlay and Rodriguez (1977) who employ the same production function as Itsumi and Moriguchi (1978) in a Keynesian model with flexible exchange rates and show that fiscal policy is able to influence real income in such a setting.

lanced trade is determined by the relationship between the growth of the labor force and that of world demand as the latter determines exports. If the rate of growth of the labor force is the smaller one, the long run steady state of the economy converges to this rate. But if world demand grows more slowly than the labor force, the long run increase in output falls short of the natural rate.

3.2.3. Limited substitutability

While the influence of the international environment in fixed-proportion models obviously stems from the bottleneck character non-competitive imports can acquire, the basic result that internal growth is not independent of international circumstances also carries over to a setting in which limited substitutability between imported and domestic factors of production is possible, as Bardhan and Lewis (1970) show.

They examine the steady state growth rate of an economy that imports intermediate goods (M) and employs them in a production function assumed to be of a Cobb-Douglas type, with domestic capital (K) and labor (L) as further arguments:

$$(3.5) \qquad Q = K^{\alpha} M^{\beta} L^{1-\alpha-\beta} .$$

Savings constitute a constant fraction of national income (defined as gross output Q minus imports) and labor grows at a constant natural rate n. The world demand for exports (X) is characterized by a price-elasticity μ and grows at a constant rate f.

$$(3.6) \qquad X = P^{\mu} e^{ft},$$

where P is the price of the domestic good in terms of the imported input, the price of which is assumed constant and set equal to unity; t is again the time index. The model is closed by assuming that trade is balanced (M=X). If all factors of production are priced at the value of their marginal product, the steady state growth rate of the capital stock becomes

$$(3.7) \qquad \frac{dK}{K} = \frac{f\beta - n\mu(1-\alpha-\beta)}{\beta - \mu(1-\alpha-\beta)} .$$

It can be observed from equation 3.7 that if the growth rate of world demand falls short of the 'natural' growth rate of the labor force ($f<n$), the steady state growth rate of the capital stock is lower than the rate of population growth ($[dK/K]<n$), while exceeding the growth rate of world demand, since it is a weighted average of f and n.

Bardhan and Lewis also show that the steady state growth rate of income varies with the elasticity of substitution between the imported and the domestic factors of production: The Cobb-Douglas production function (equation 3.5) is replaced by a more general formulation. The weighted average of the growth rate of world demand and the rate of population growth (which determines the steady state income growth rate) shifts towards the more slowly growing factor if the elasticity of substitution is less than unity. If $f<n$ in such a constellation, the "sluggish pace of export demand would have had a larger depressing impact on the equilibrium rate of growth" (Bardhan and Lewis 1970, p.377, fn 1). For this reason, it is of interest to examine the importance of imported factors of production in macroeconomic production functions empirically, as *the internal growth rate of the importing country varies with the degree of substitutability for imported factors of production*.

Before turning to studies that have tried to model imports as inputs in empirical analyses we take a closer look at the 'standard' import demand function generally encountered in empirical studies and problems associated with its implementation.[9]

3.3. The standard import function, foreign exchange availability, and disaggregate import analyses

3.3.1. The standard import function

The traditional approach to modelling import demand behavior for empirical analyses is the 'import-substitutes model', which differs from the pure trade-theoretical approach (Heckscher-Ohlin-Samuelson model) since it takes into account the imperfect substitutability between traded and domestically produced commodities (Armington assumption).[10] It is derived from microeconomic considerations: A utility function is postulated in which both domestic and imported goods are arguments. A maximization of the utility function subject to a given income (Y) constraint determines the amount of imports (M^d) depending on the degree of substitutability and the relative price ratio between the domestic and the imported good.

9 For extensions of the analysis of open economies with imported inputs see especially Herberg et al. (1982), Jones and Purvis (1983), Obstfeld (1980), Sanyal and Jones (1982), Schmid (1982), Van Bochove (1982), and Ziesemer (1987).

10 Compare Khan and Knight (1986, p.12) for the implications of the perfect substitutes assumption. The introduction of the imperfect substitutes assumption is due to Armington (1969), who distinguishes products by kind and place of production.

$$(3.8) \qquad M^d = M^d\left[Y, \frac{P_D}{P_M{}'E}\right],$$

where P_D and P_M' represent the price of a domestic variable (most often the GDP deflator or the wholesale price index is chosen) and the border price of imports (in foreign currency); E stands for the exchange rate of the domestic per one unit of foreign currency. This functional relationship has been employed in numerous models trying to explore the determinants of import-behavior for both developed and developing countries. For the latter, the appropriateness of such a function was doubted for a long time, largely because developing country governments often play a discretionary role both in trade policy (via exchange controls, quantitative and tariff restrictions) and as the actual unit which decides type and quantity of specific import goods. Since non-market forces were thought to govern the import behavior in developing countries, the inclusion of a relative price term was neglected for a long time.

Following Houthakker and Maggee (1969), Khan (1974) examines the appropriateness of an equation like 3.8 for developing countries. He employs a modified version of the equation to explore import demand determinants of fifteen developing economies. The estimated import price elasticities are found to be relatively high and indicate, according to Khan (1974, p.687), that relative prices have a significant effect on the imports of developing countries. Postulating import functions of the described form has become commonplace also for developing countries.[11]

Several problems arise when modelling import behavior in the above functional form. The first relates to treating imports as consumption goods only, thus ignoring production linkages.[12] If domestic output can itself be expressed (among other arguments) as a function of the level of imports, the direction of causation between changes in imports and output is ambiguous. Imports determine output and vice versa. A direct estimation of an import demand function like 3.8 then produces a simultaneity bias.[13]

The other problems are related to the modelling of import demand functions in general, irrespective of whether they describe consumer demand or conditional demands for factors of production. Therefore, they are of importance for the empirical analysis which we intend to carry

---

11 Compare, e.g., Khan (1975) and Melo and Vogt (1984).

12 As will be pointed out below, there is one possibility of describing a derived demand function for imports of cost minimizing agents solely by a relative price variable and domestic income.

13 Compare Goldsbrough and Zaidi (1986, p.172) on this issue.

out. One problem is concerned with the neglect of the external budget restriction, namely that the import value in every period is equal to export revenues plus the net change in foreign assets. The second relates to the aggregation of inhomogeneous import categories.

3.3.2. Foreign exchange and the external budget restriction

In any given period, import expenditures balance the sum of export revenues, net interest receipts, the net inflow of direct investment, net external borrowings, and transfers minus an increase in the foreign exchange reserves. Estimating an equation like 3.8 assumes that import behavior is independent of foreign exchange availability. It is postulated that, in the balance of payments equation, imports are determined by private and government demand decisions to which other components of the balance of payments have to adjust.[14] The 1980s have proven, however, that a number of developing countries were not able to finance desired imports since the availability of international credit was restricted and export revenues could not easily be increased to make up for the lower capital inflow. With respect to this period Moran (1989, p.279) asserts that "although the traditional import model, which links imports to domestic output and relative import prices, has worked well for industrial countries that are unconstrained by foreign exchange, it has not proven useful in explaining the recent slump in the imports of developing countries which are short of foreign exchange".

If desired imports cannot be financed with available foreign exchange, several adjustments can occur. One is the devaluation of the home currency, which alters the relative price ratio of traded to nontraded goods and eliminates the excess demand for imports as the domestic currency price of imports relative to the price of domestically produced import substitutes is raised.

The other possibility is a 'forced adjustment' of imports which operates through the imposition of trade policy measures. These include raising tariff levels, imposing quantitative restrictions, or making use of other non-tariff barriers such as quality controls or administrative obstacles. The internal price of imported goods (P_M) is equal to the product of the border price (P_M') and the combined effect of all tariff and non-tariff measures $(1+t_i)$:

$$(3.9) \qquad P_M = P_M' \, (1+t_i)$$

---

14 The foreign exchange availability is also of little or no importance if domestic and foreign wealth are perfect substitutes, as Lopez and Thomas (1988, p.42) point out. If foreign debt or imports can be repayed by domestic assets or income, no attention would have to be given to a foreign exchange constraint.

If the import reduction is achieved solely through the use of ad valorem tariffs, data on these nominal tariffs can be used to calculate the internal price of the imported goods. The tariff-corrected price replaces the border price in the import demand function (3.8). Similarly, if the authorities auction import licenses under quantitatively regulated trade, the prices of the licenses help to determine the price faced by domestic importers.[15] But developing countries do not merely rely on tariffs to curb import demand; auctioning of licenses combined with respective price reports is also seldom encountered.

One might think that calculations of the effective rate of protection (ERP) help to determine the combined effect of all protectionist policies needed in empirical analyses of import demand functions. ERPs are most often computed by comparing the prices of domestically produced goods to the world market prices (or border prices) of their traded counterparts.[16] According to the law of one price, the domestic price should equal the border price plus the combined effect of tariff and non-tariff barriers, taking into account subsidies and trade taxes on imported inputs. One important theoretical caveat arises: The effective rate of protection measures are based on the assumption that the domestically produced goods and the imported ones are *homogeneous* products and therefore perfect substitutes. This is in obvious contradiction to the Armington assumption since imperfect substitutability implies that the law of one price does not hold. Empirical evidence strongly contradicts the validity of the law of one price.[17] "In short, the 'law of one price' does *not* seem to hold either across or within countries, except perhaps for standard commodities such as wheat or copper that are sold on international commodity exchanges" (Goldstein and Khan 1985, p.1045).

Accordingly, import demand models do not employ calculations of the effective rate of protection based on price comparisons. But if the impact of import restrictions were neglected altogether in the estimating equation, this would cause a bias of the price elasticity estimate unless trade restrictions are completely uncorrelated with other exogenous variables. Alternative measures of the effect of protectionist policies are used. We summarize the main approaches in Table 3.1.

---

15 Compare Eggerstedt et al. (1987), Greenaway (1983), Hamilton (1984), and Siebert (1986) on the economic effects of quotas and allocation mechanisms.

16 See, e.g., Zietz and Valdes (1986).

17 See Goldstein and Khan (1985, p.1045) for a list of studies that examine and negate the validity of the law of one price.

Table 3.1:
Classification of studies which include the external budget restriction

	Inclusion of a protection proxy (e.g. nominal tariff)	Government reaction function	
		Premium rationing	Fix-price rationing
examples of studies	Bautista (1980) Conway (1986)	Dutta (1964) Faini et al. (1988) Lopez and Thomas (1988) Moran (1989) Turnovsky (1968)	Khan and Knight (1986) Marquez (1984, 1985)

Some authors include a protection proxy such as the nominal tariff rate or the effective tax on imports (import revenues divided by the import bill). These measures do not include the effect of non-tariff barriers to trade, but it is assumed that quantitative restrictions or other non-tariff measures are tightened and loosened concurrently with changes in the included protection proxy.

In the extreme case in which the effect of quantitative restrictions is proportionate to $(1+t_i)$ in equation 3.9, the percentage change in the relative price of the domestically produced good via the traded good would be unaffected. The effect of the quantitative restrictions is then picked up in a constant term included in an estimation of the import demand function 3.8. If the included and the excluded relative price effect are correlated, on the other hand, this does have an influence on the parameter estimates. The estimate of the price elasticity of import demand would be biased.[18]

A different approach specifies a 'government reaction function' in which the use of policy measures is endogenized and specified to be dependent on an exogenous variable. It postulates that a private sector demand function for imports exists, but "that the supply of imports is really the supply of foreign exchange made available to purchase imports" (Faini et al. 1988, p.11). If the desired demand for imports at the world price is not compatible with the foreign exchange reserves, the government restricts imports so as to equate available foreign exchange with demand. The approach consequently suggests that a foreign exchange variable is included directly in the import function.[19]

---

18 Compare Hemphill (1974), pp.637-638.

19 See Faini et al. (1988) and Moran (1989) for a discussion of various import models that introduce the foreign exchange restriction.

Two protectionist reactions to a foreign exchange shortage are employed in empirical studies of import demand. One is 'premium-rationing', where the imposition of protectionist policies raises the domestic price of the imported goods. The other is termed 'fix-price rationing' to underline that the policies do not alter the price.[20]

Premium-rationing measures include all tariff and non-tariff barriers that have an influence on the price of the good as encountered by the importer. They include an increase of the nominal tariff rate, imposition of surcharges or other fiscal charges, leverage of taxes on foreign exchange transactions, a rise in the price paid by the importer for quota allotments, or any other additional costs associated with obtaining the import license. This effect can be accounted for in import demand models by including a foreign exchange variable in the estimating equation. Different measures were put to use in the literature. Dutta (1964), for example, includes the ratio of foreign exchange holdings in period t to private sector imports of the previous period. Turnovsky (1968) assumes foreign assets and export proceeds to represent the relevant foreign exchange variable, while Lopez and Thomas (1988) include both the value of exports and debt in their import demand equation. Hemphill (1974) omits both price and income measures and only uses current and lagged foreign exchange receipts in the estimated import equation. He views this more or less as a reduced-form equation, since the given amount of foreign exchange determines price and income measures, which in turn result in an observed import volume (Hemphill 1974, p.642).[21]

Fix-price rationing measures, on the other hand, are defined as ones that do not affect the internal price of imported goods. One might envisage import allowances to be distributed among the demanders in a predetermined way, for example in proportion to desired demand. The importers do not have to pay for the licenses and no auctioning or resale system exists. Rather than assuming that the foreign exchange shortage pushes up the domestic price of imported goods, the time needed for the actual import level to adjust to the desired one is influenced by the availability of foreign exchange. If the supply of foreign exchange is large, no additional quantitative controls have to be imposed and import demand is satisfied with delivery lags accounting for a difference between desired and actual levels. If foreign exchange is scarce, on the other hand, the authorities extend the period until a license application is decided on. It is assumed that the costs of imports do not change, while the time span to obtain desired imports is prolonged. Khan and

---

20 Compare Dervis et al. (1981).

21 Winters (1987) follows Hemphill (1974) in linking the import decision to financial variables. He develops a model in which the import decision is based on the maximization of an intertemporal import sub-utility function, independent of any other variable. The developing country can incur debt in order to finance a higher stream of imports today, but the intertemporal budget restriction that all imports have to be paid for in the long run is binding.

Knight (1986) and Marquez (1984, 1985) include such a mechanism in their import demand models. The speed of adjustment of actual to desired import quantities is a function of a foreign exchange variable.

Both of the above protectionist responses are of relevance. It has been observed that tariff levels, surcharges, and foreign exchange taxes were raised in a lot of indebted developing countries in order to curb import demand in the 1980s. Auctioning or resale of licenses sometimes occurs and rent seeking activities can also bid up the price of the imported good in the domestic market. Some influence of the foreign exchange availability on the adjustment time is likely as well, since the bureaucratic procedure of granting the allowances might well be stretched.

The strong assumption underlying the variable speed of adjustment models is that the user cost of the imported good is unaffected by the foreign exchange shortage. But opportunity costs of importing are raised in this case, too, if the domestic agent (consumer or producer) is aware of the prolonged duration until the application is handled. The utility of consuming the import good, and the profitability of an investment with it, are discounted. The longer adjustment time has an effect on the user cost as well.

It is difficult to include both of the government reaction functions described above in a single import demand equation. An empirical formulation would have to be chosen, which incorporates a variable speed of adjustment of import demand and an effect of the foreign exchange availability on the domestic price of the imported good. Two foreign exchange variables to represent these two effects would need to be included in an estimation. Multicollinearity problems arise even if two different foreign exchange variables are chosen since they are likely to be highly correlated. As trade barriers causing a rise in the internal price of the imported good cover a broader spectrum of protectionist instruments than those influencing the adjustment time to desired import levels, we favor the inclusion of a foreign exchange variable that represents the impact of policies on the internal price of imported goods.

3.3.3. Disaggregate import analyses

Equation 3.8 describes the demand for *aggregate* imports and does not distinguish between different product classes. "If the effect of the determining variables is exactly the same as between the aggregate and disaggregated groups, or if the relationship between the components and aggregate explanatory variables is a stable one, then one can be indifferent between the aggregate and disaggregate equations" (Goldstein and Khan 1985, p.1070). But either one of these preconditions is not likely to hold so that some studies have explored income- and price-elasticities on a more detailed level. The Armington assumption of imperfect substitutability between imported and domestically produced goods is applied to more detailed product specifications. Evidence collected for the industrial countries suggests that price elasticities are higher for manufactures than for non-manufactures.[22] Disaggregate import demand analyses for developing countries is scarce. We summarize briefly results of some studies below.

Sarmad (1989) suggests that both price- and income-elasticities are likely to vary with the income level and substitution efforts of developing countries in specific industries. While he finds price-elasticities for Pakistan to be highest for machinery and transport equipment (which he attributes to the Pakistani substitution policies for these goods), Khan (1975) and Melo and Vogt (1984) attain their lowest price-elasticities in the Venezuelan import demand for precisely the same import category.[23]

The studies by Mirakhor and Montiel (1987) and Salas (1982) do not distinguish commodity classes by the nature of the traded commodities (manufactures versus non-manufactures), but by the end-use in the importing country, namely capital, intermediate, and consumer goods. They attempt to model the production linkages between imports and domestic output of the importing country more precisely, but do not specify a formal production function. In his estimating equation, Salas (1982) finds that Mexican imports of intermediate goods show a higher price elasticity than capital goods, which he (like Sarmad 1989) relates to the almost complete import substitution of intermediate goods in Mexico.

Mirakhor and Montiel (1987) explore the determinants of import demand for capital and intermediate goods for fourty-seven developing countries. In both equations, they include the terms of trade, which are assumed to have a positive impact on imports since they raise expectations of future output increases and ease liquidity constraints through improved cash flows, real output, the real exchange rate as an important determinant of the relative price between imported and domestically produced goods, and the real interest rate.[24] A dynamic adjustment process is

---

22 Compare Goldstein and Khan (1985, p.1070 and Table 4.4 on p.1086).

23 Compare also Ghose et al. (1986), who model a sequential determination of Indian import demand.

24 In the capital goods equation, the real interest rate is included as an exogenous variable because it is a major determinant of the user cost of capital. It is incorporated in the equation for imported intermediate goods since it influences the demand for inventories and thus indirectly for imports of intermediate goods.

not included in the estimating functions,[25] although empirical evidence of import demand functions strongly suggests that considerable time lags for quantity adjustments exist.[26]

Real output is determined to be the most important exogenous variable in both import demand equations estimated by Mirakhor and Montiel (1987). While relatively strong support is found for the view that the terms of trade variable influences the decision to import capital goods, this is not true for imported intermediate goods. This result has been anticipated by the authors, as expectations about future output increases motivate primarily investment and hence capital goods imports. The other two variables perform poorly. In only about half of the estimations does the real exchange rate variable carry the correct sign. This might be due to the fact that Mirakhor and Montiel do not construct a variable that directly links the price of imported goods to their domestic substitutes. The real interest rate appears with the expected negative sign in only eighteen regressions for capital goods (none significant) and in about half of the estimations for imported intermediate goods (four significant).[27]

All of the above mentioned studies conclude that the impact of exogenous variables on import demand differs across commodity classes.

3.4. Empirical studies of imports as factors of production in developing countries

We now turn to models which explicitly treat imports as inputs. The first sub-section is concerned with studies that posit a specific value for the elasticity of substitution. The second subsection then examines import models which are flexible in this respect.

---

25 This might be due to the short period of estimation (1970-1981) which leaves the authors with only twelve observations and seven degrees of freedom.

26 Compare the survey by Goldstein and Khan (1985, pp.1066-1069 and their Table 4.2. on pp.1080-1081).

27 The poor performance of the real interest rate variable might be due to restrictive financial policies which a lot of developing countries pursue. For descriptions and analyses of interest rate policies in developing countries see Arrieta (1988), IMF (1983), and Khatkhate (1988).

3.4.1. Models positing a specific elasticity of substitution for imported inputs

3.4.1.1. Trade-gap models

Most authors who examine the production linkages between imports and growth assume certain values for the elasticity of substitution between imported and domestic factors of production. Multi-equation models linking the foreign trade and domestic sector while explicitly modelling imports as requirements in output growth generally relate current real imports of capital goods to total investment (current and/or lagged). Such a formulation is used by Bacha (1986) and the World Bank SimLink (Simulation of Trade Linkages) model (Hicks 1976), as well as for the projections model of the Inter-American Development Bank (1984) and Park (1985). These models use a specification method which stems from the structural approach viewing domestic and imported capital goods as complements in the production process.[28]

Several objections have to be raised when such a demand specification for imported capital goods is used:

(a) Although a theoretical analysis in which imported and domestic means of production are perfect complements is justified as describing one out of several possible relationships, the a priori assumption of perfect complementarity in empirical analyses is questionable. None of the studies mentioned above attempts to estimate the elasticity of substitution; prices and other possibly relevant variables are omitted.

(b) Even if the assumption of complementarity holds, the use of total investment as an explanatory variable biases the results. Consider the simple equation used by Park (1985, p.841):

$$(3.10)\quad M_{CAP} = \beta_1 + \beta_2 I_t + \beta_3 M_{Cap,t-1} + u_t$$

---

28 Other models in the tradition of the two-gap approach also assume complementarity in production of imported and domestically produced machinery and equipment, but use either no estimation techniques (Farzin 1988) or derive different reduced form equations (El-Shibly and Thirlwall 1981, Fajana 1979, Voivodas 1973). See Behrman (1971) for a critique of the estimation of two-gap models.

Zaidi (1988) employs the complementarity assumption in a rationing model of foreign exchange. Marquez and Pauly (1987) use an indirect approach modelling the dependence on imported capital goods. They relate the change in the capital stock to available foreign exchange reserves, postulating that the latter cause imports of capital goods to be curtailed which in turn depresses aggregate investment. Similary, UNCTAD's System for Interlinked Global Modelling and Analysis (SIGMA) links gross fixed capital formation to the availability of imported capital goods (Richtering 1989, p.528).

where M_{CAP} and I_t represent capital goods imports and investment. $M_{Cap,t-1}$ is the lagged endogenous variable, $ß_i$ are parameters, and u_t is the error term. But total investment itself consists of domestically produced capital goods and imported capital goods. If the error term in equation 3.10 jumps up, it directly increases the imports of capital goods, which in turn increase investment. Since investment is an exogenous variable in equation 3.10, the disturbance u_t and this regressor are positively correlated. Estimation with ordinary least squares produces an (even asymptotically) upward biased estimate of $ß_2$.[29]

Multi-equation models designed for simulation exercises assume a complementary production technology for imported intermediate goods too.[30] They are linked to scale variables such as manufacturing value added or gross output. Again, the a priori assumption that imported supplies cannot be substituted for by domestic factors of production is questionable. Relevant exogenous variables might be omitted in the demand equations for intermediate imports.

3.4.1.2. Models positing a limited substitutability

Some authors relax the perfect complementarity assumption and posit a limited, but predetermined substitutability instead. Bergsten et al. (1985) estimate a Cobb-Douglas function for Latin American countries in which imports (the average of current and lagged values) enter beside the fixed capital stock, labor, and time. They thereby restrict the elasticity of substitution between imports and domestic factors to 1. The calculated elasticities of output with respect to imports range from 0 for Argentina to around .14 for Brazil, .23 for Mexico, and .27 for Chile. These figures are likely to contain a bias. By incorporating both the fixed capital stock (including current investment expenditure) and total imports of the same year, imported capital goods are counted twice, since they are included in the expenditure data on gross fixed capital formation.[31]

---

29 On the bias in these estimations compare also Weisskopf (1979, p.63, fn 16). Investment as an exogenous variable appears also in other studies, which include additional exogenous variables influencing the demand for foreign capital goods. The correlation problem mentioned above is present in these studies too. See Rashid (1984) and Salas (1982).

30 Compare Farzin (1988), Fritsch (1988), Hicks (1976), Inter-American Development Bank (1984), Oberhänsli (1981), and Park (1985).

31 A similar framework is employed by Cardoso and Fishlow (1989a), who regress growth rates of output of Latin American countries on growth rates of employment, imports (representing in their view the foreign exchange restraint), exports (standing for positive externalities associated with outward-oriented activities) and the share of investment in GDP. This is termed an 'augmented' neoclassical production function. The criticism raised above (bias due to double-counting of imported capital goods) applies to this study as well. They find that imports constitute an independent influence on output growth; according to their estimates, a 1% growth in imports accounted for about .15% of growth in GDP of Latin America in the 1950-1980 period.

Khan and Knight (1986, 1988) specify an export production function of developing countries with imported inputs and value added by domestic factors of production as arguments. The factors are combined in a Cobb-Dougas production function; the elasticity of substitution both between capital and imported inputs and between labor and imported inputs is thereby restricted to 1. They estimate the elasticity of the volume of gross exports with respect to imported inputs for 34 developing countries at .52, a fairly large estimate as this implies that exports are highly import-intensive and that imported inputs account for more than the combined contribution of domestic factors of production in total export production.[32]

3.4.2. Models that allow estimation of the elasticity of substitution

3.4.2.1. CES functions including imports

As observed above, the majority of empirical studies establishing a link between imports and domestic output, assume a certain functional form which restricts the elasticity of substitution between inputs to certain values (0 for Leontief technology, 1 in the Cobb-Douglas function). In order to analyze the role imported capital and intermediate goods play in developing economies, it is important to specify a much more flexible functional form, in which elasticities of substitution between different input pairs can take on a wide range of values.

Consider a production function of the general form

$$(3.11) \quad Q = f(V, M),$$

with $V = V(V_1, \ldots, V_m)$ and $M = M(M_1, \ldots, M_n)$, where Q is gross output, V represents domestic factors of production, and M stand for a vector of imported inputs. The elasticities of substitution between the different pairs of inputs are likely to vary, some input combinations showing substitutability relations ($\sigma \geq 1$), while others show a complementarity behavior in the production process ($0 \leq \sigma < 1$).

The Constant Elasticity of Substitution (CES) production function allows for such a flexible form. The original CES function (Arrow et al. 1961) was developed for two factors of production. The extension to a multi-input CES function (Mukerji 1963) with, correspondingly, mul-

32 Computable General Equilibrium (CGE) models of developing countries also include imports as means of production. They are, however, not estimated econometrically and set production function parameters. Compare Dervis et al. (1982) Dick et al. (1982a, 1982b).

tiple elasticities of substitution showed that restrictions have to be imposed on the values of these elasticities.[33]

Sato (1967) developed the 'nested' version of the CES function, employing the concept of separability between variables. The concept is of direct importance in a multi-input production model, since it concerns the structural property which may "permit economic analysis to be carried out in terms of subsets of the total set of possible variables, in stages, or with consistent aggregates of variables" (Fuss et al. 1978, p.221). Separability implies "uniform or invariant behavior of certain economic quantities" (Kaneda 1982, p.102). If a production function is specified as

$$(3.12) \qquad Q = f(V_1, V_2, M),$$

the two inputs V_1 and V_2 are said to be weakly functionally separable from the input factor M, if the marginal rate of substitution between V_1 and V_2 is independent of the level of the factor M:

$$(3.13) \qquad \frac{\partial(Q_{V1}/Q_{V2})}{\partial M} = 0.$$

Q_{V1} and Q_{V2} represent the marginal productivities of the factors V_1 and V_2. If V_1 and V_2 represent capital and labor and M stands for total imports, functional separability between these inputs would require that an increase in the quantity of imports leads to a proportional rise in the marginal productivities of capital and labor.[34] If functional separability holds, capital and labor can be nested and the combination of the two joined together in the aggregate production function. Profit maximization (for given total costs) or cost minimization (for a given output level) can therefore take place in two distinct stages: First, the optimal factor use of imports and the aggregate of the domestic factors of production is determined. Second, the allocation of capital and labor to form the domestic factor combination is derived. Using the concept of separability with CES functions permits different elasticities of substitution beween different pairs of inputs -

---

33 It would require that either the partial elasticities of substitution between all pairs of inputs are constant and identical or that the ratios of substitution elasticities are fixed.

34 The concept of weak functional separability in the three-input case is equivalent to the concept of Hicks-neutral technological change in the two input case (compare Kaneda 1982, p.102). See Berndt and Christensen (1973) for an explanation of the separability concept.

something which cannot be achieved using a multi-input CES function without nesting.[35] The concept of nested production functions has achieved widespread use in economic model building.[36]

A number of empirical studies have employed CES production functions to estimate elasticities of substitution between imports and domestic factors of production for developing countries. Some employ the concept of separability if more than two variables are included, others use a two-variable specification. Table 3.2 summarizes the empirical findings of some studies which all derive estimation functions assuming that firms choose factor inputs in an optimal way. Michalopoulos (1975) and Marquez (1984) incorporate a dynamic adjustment process so that short run elasticities can be distinguished from long run elasticities. Both studies support the finding of other empirical studies that adjustment of imports to their desired level takes considerable time. The estimates of the first three studies shown in Table 3.2 suggest that imported intermediate goods are complements to domestic factors of production in developing countries. Michalopoulos (1975) uses a nested production function, separating a domestically produced capital stock from one that is derived solely from imported machinery. The short run substitutability is found to be low (.22), while the long run elasticity of substitution is much higher (2.263). He employs a partial adjustment mechanism in his empirical formulation and estimates the speed of adjustment (λ) at .1, an estimate not significantly different from zero.[37] Consequently, his derived estimate of the long run elasticity of substitution between domestic and imported capital goods is not significantly different from zero and questions have to be raised whether this estimate really indicates long run substitution possibilities in the Argentinian economy. The same caveat also holds for the long run elasticity estimate reported in Marquez (1984).

---

35 If the separability assumption is applied to CES functions, strong separability has to hold in order to allow for modelling a function $Q = f\{J(V_1,..,V_i) + G(V_{i+1},..,V_n)\}$. See Sato (1967) and Berndt and Christensen (1973).

36 It has been applied to the agricultural (Kaneda 1982) and manufacturing sector (Sato 1967), and in the analysis of the labor market (Costrell et al. 1986). But most often it is used to take account of imported factors of production (e.g. Conway 1986, Marquez 1984 and 1985, Taylor 1979).

37 The coefficient of the lagged endogenous variable (1-λ) is .9009 in Michalopoulos'(1975, p.350) estimation with a standard deviation of .08681. It is not significantly different from one which would imply that no adjustment takes place. Marquez (1984, 1985) also examines the relationship between an imported and a domestically produced capital stock. But he derives his estimating equation on the hypothesis that σ_k, the elasticity of substitution between the two capital stocks, is very small (compare Marquez 1985, p.47).

Table 3.2:
Estimated elasticities of substitution between imported and domestic factors of production in developing countries, selected studies[1]

Study	Countries	Factor pair	Short-run elasticity	Long-run elasticity
Conway (1986)	Turkey	domestic value added, imported intermediate goods	.	.66
Marquez (1984)	developing countries	capital stock, imported oil	.0567	.7351
Marquez and Pauly (1987)	developing countries	capital stock, imported oil	.	.73
Michalopoulos (1975)	Argentina	domestic capital stock, imported capital stock	.2242	2.263

1 **All studies employ CES-functions to estimate the factor-pair elasticities of substitution. Conway (1986), Marquez (1984), and Michalopolous (1975) employ nested CES functions. Conway (1986) and Marquez and Pauly (1987) do not distinguish between short- and long-run elasticities.**

3.4.2.2. The translog approach

One major deficiency associated with estimation of multi-input nested CES functions is that separability between different pairs of inputs has to be assumed in order to enable the researcher to conduct consecutive estimations at different stages of the production framework. Conway (1986), for example, assumes that total output can be expressed as a function of domestic value added and intermediate inputs, whereas the second level expresses domestic value added as a function of the installed capital stock and labor. This functional form implies that imported intermediate goods are separable from capital and labor, restricting the elasticity of substitution between imported intermediate goods and labor to equality with the elasticity of substitution between imported intermediate goods and capital.

One possibility of avoiding such a separability assumption while still examining multi-input production functions is offered by the transcendental logarithmic (translog) function which is due to Christensen et al. (1971). It expresses the logarithm of output as a quadratic function of the logarithm of inputs:[38]

38 The translog function is a second-order approximation to a generalized cost or production function. For a discussion of the translog function see Hamermesh and Grant (1979) and Kaneda (1982).

$$(3.14) \qquad \ln Q = \ln\alpha_0 + \sum_i \alpha_i \ln V_i + \frac{1}{2} \sum_i \sum_j \beta_{ij} \ln V_i \ln V_j,$$

where Q stands for output, α_0, α_i, and β_{ij} are parameters, and V_i and V_j are inputs. If all $\beta_{ij} = 0$, the form reduces to a multi-input Cobb-Douglas function. "The quadratic terms can be regarded as amendments to the Cobb-Douglas assumption of unitary elasticity of substitution" (Kaneda 1982, p.100). The β_{ij} can be used to calculate the Allen Partial Elasticities of Substitution (AES) between all input pairs. While the β_{ij} are assumed constant in empirical modelling, the partial elasticities of substitution derived from them are variable. The transcendental production function allows testing the separability between input pairs. Separability of input factors V_1 and V_2 versus a third factor V_3 would imply that $\sigma_{13} = \sigma_{23}$, where σ_{ij} is the AES between factors i and j.

Rather than employing the translog production function (equation 3.14), most empirical studies analyze the translog cost-function:

$$(3.15) \qquad \ln C = \ln\gamma_0 + \sum_i \gamma_i \ln W_i + \frac{1}{2} \sum_i \sum_j j_{ij} \ln W_i \ln W_j,$$

with C as the minimum cost associated with a given output level Q in the production function, γ_0, γ_i, and j_{ij} as parameters, and W_i and W_j as factor prices. Applying Shephard's Lemma, convenient cost-share equations for each factor can be derived from this function which are the basis for the estimating equations.[39]

Table 3.3 summarizes the results of studies that employ such a translog cost function to analyze whether imports and domestic factors are complements or substitutes. Since the derived AES vary from observation to observation, studies generally conclude only whether input pairs show complementarity (AES $\sigma_{ij} < 0$) or substitutability (AES $\sigma_{ij} > 0$) relations. As this approach to deriving import demand functions for developing countries is rare, both estimates for developing and industrialized countries are included in the table.[40] All studies use total imports and do not disaggregate into intermediate and capital goods. Burgress (1974a), Mohabbat and Desai (1983), and Mohabbat et al. (1984) conclude that separability of total imports from domestic capital and labor in their country studies is not supported by the data, which implies that $\sigma_{KM} \neq \sigma_{LM}$ (the AES between imports and capital is different from the AES between imports and labor). The result may stem from the heterogeneous nature of imports since machinery is used for capital

39 Compare, e.g., Aw and Roberts (1985), Mohabbat and Desai (1983), and Mohabbat et al. (1984).

40 For further studies applying the translog cost and production functions to analyze import demand behavior see Appelbaum and Kohli (1979), Burgress (1974b), and Kohli (1978).

Table 3.3:
Translog estimations of import substitutability and complementarity in production, selected studies[1]

Study	Country	Factor pair	Factor relation
Aw and Roberts (1985)	United States	total imports from newly industrialized countries, labor	complements
		total imports from newly industrialized countries, capital	substitutes
		total imports from developed countries, labor	substitutes
		total imports from developed countries, capital	substitutes
Burgress (1974a)	United States	total imports, labor	substitutes
		total imports, capital	complements
Mohabbat and Desai (1983)	South Korea	total imports, labor	substitutes
		total imports, capital	substitutes
Mohabbat et al. (1984)	India	total imports, labor	(not significant)
		total imports, capital	substitutes

1 All studies employ translog-cost functions to calculate Allen partial elasticites of substitution between factor pairs.

formation while intermediate goods are used to uphold current output directly. If a separation of imports from capital and labor were possible, this would justify a functional form as depicted in equation 3.8, where domestic value added (income) and relative prices appear as exogenous variables. But as pointed out above, price responses of disaggregate import classes in standard or extended import demand functions have been found to differ significantly from each other, which questions their aggregation into one composite factor.[41]

3.4.2.3. Derived import demands from CES and translog functions: a comparison

While the multi-input transcendental production or cost function has the advantage that separability relations between pairs of factor inputs can be tested, it has major disadvantages compared to the CES function.

41 In the case of developing countries see especially the above mentioned study by Mirakhor and Montiel (1987).

First, the property of 'self-duality' does not hold for the translog production and cost function. A translog cost function may be homogeneous of degree one without implying that the corresponding production function is homogeneous of degree one (Kaneda 1982, p.101). The property of self-duality of CES functions is of major importance for multi-equation models, since it permits the inclusion of derived demand functions for factor inputs as well as the corresponding dual cost functions.[42]

The second and more important disadvantage of translog functions compared to a nested CES framework is that the translog approach does not allow for a straightforward modelling of a dynamic version of it, since lagged variables cannot be simply added to the equation (Costrell et al. 1986, p.782).[43] This is a serious impediment when time series data are analyzed and even more important if imports are included in the production function, as empirical studies strongly suggest that imports require considerable time to adjust to desired levels.[44]

Further, it is much easier to introduce an additional variable such as a foreign exchange constraint in derived import demand functions from CES production functions than in the translog case. Therefore, the practical applicability, especially since we want to include the results in a multi-equation model, suggests giving preference to CES functions. The result from translog separability tests that *total imports* appear not to be separable from domestic factors (Burgress 1974a, Mohabbat and Desai 1983, and Mohabbat et al. 1984) should nevertheless be kept in mind.

3.5. Concluding remarks

In this chapter we have dealt with the empirical modelling of import functions. Some issues were discussed which are relevant to our aim of estimating elasticities of substitution between imported and domestic factors of production:

(a) Standard import demand functions that link imports to income and a relative price term are not appropriate in the case of imported inputs. Causality runs from income to imports and vice versa in this case, which implies that regression results are biased.

---

42 See Sato (1967), Uzawa (1962), and Appendix B.

43 Compare also Hamermesh and Grant (1979, p.521) and Shiells (1989, p.126). See Pindyck and Rotemberg (1983a, 1983b) for the specification of a dynamic translog input requirement function.

44 See the survey on import demand functions by Goldstein and Khan (1985). The study by Michalopoulos (1975) included in Table 3.2, for example, finds that a mean adjustment time of the actual to the desired ratio of the domestic to imported capital stock takes about nine years in Argentina.

(b) Standard, extended, and derived factor demand functions for imports are based on the Armington assumption of imperfect substitutability between domestic and imported goods. We pointed out that the inclusion of effective rates of protection derived from a price comparison between domestically produced and imported goods (taking into account subsidies and tariffs on imported inputs) is not compatible with the Armington assumption.

(c) The effect of foreign exchange shortages can be incorporated in import demand models either by using a protection proxy or by specifying a trade policy reaction function of the government. Two possible reaction functions exist: One attempts to incorporate the effect of protectionist policies on the internal price of imported goods, the other assumes a functional relationship between available foreign exchange and the adjustment time of import demand. It is difficult to simultaneously incorporate both of these functional relationships in one estimating equation. We favored the inclusion of the price-influencing reaction function since it covers a broader spectrum of actual protectionist policies.

(d) It is important to study disaggregate import functions. Studies that examine disaggregate import demand models find that price elasticities vary across import categories.

(e) Most empirical studies of imports as inputs posit a specific elasticity of substitution between imported and domestic factors of production and do not test it. Two models are appropriate in this respect: the translog approach and the nested CES production function. Both allow us to specify imports as inputs in a multi-factor production function without restricting the elasticities of substitution between input pairs. The nested CES framework has one major advantage compared to the translog approach since it allows for inclusion of dynamic adjustments in a straightforward way.

4. Macroeconomic production functions and elasticities of substitution between imported and domestic factors of production

4.1. Introduction

In order to quantify the importance of foreign-made inputs in developing countries, the elasticity of substitution in the production process has to be *estimated* instead of being assumed on *a priori* grounds. As was pointed out in the previous chapter, little empirical work on this subject has been carried out. Marquez (1984, 1985) specifies a production function similar to the one outlined here, but he treats the non-oil developing countries as an aggregate. But we observed in Chapter 2 that the group is not homogeneous with respect to the stage of development, the level of industrialization, and development strategies of the individual countries. A further inquiry into the technical production relations on an individual country basis is important on that account.

This chapter outlines a macroeconomic production framework which includes both imported and domestic factors of production. We choose a nested constant elasticity of substitution production function to represent the functional form. Imported intermediate inputs, labor, and two different capital stocks are specified as arguments in the production function. One capital stock is made up of domestically produced investment goods while the second consists of imported machinery and equipment. Assuming that individual firms minimize production costs subject to a given production constraint, several equations are derived which are then estimated in the next chapter. After discussing several features of the elasticities of substitution in Section 2, the third section introduces the system of production functions. Reduced form equations suitable for parameter estimation are derived from the first order conditions of a cost minimum.

4.2. Aggregate production functions and the elasticity of substitution

4.2.1. The aggregation problem

In theoretical macroeconomic models, output is most often assumed to be a homogeneous good which can be either invested, exported, or consumed. The assumption paves the way for aggregating the production technology and the technological relationships between inputs and total output in one single production function. In reality, however, several of the underlying conditions that would make the macro-production function describe real economic circumstances in a precise way are unlikely to hold. First of all, no economy produces only one single, completely homogeneous commodity. And even if a country specializes to a large extent on the production of one (e.g. agricultural) output, it is hardly possible to satisfy both investment and consumption demand with this one product.

Additionally, the factors of production - especially capital goods - cannot be assumed to be homogeneous across sectors and industries of an economy.[1] Each micro-production function is associated with its own technique, which is embodied in its capital stock (Sato 1975, p.4). If all micro-functions are known, it is possible to aggregate them 'one by one' and describe the production of the economy in the following form:

$$(4.1) \qquad Y = f(K_1, \ldots, K_n, L),$$

where Y represents domestic value added and K_1 to K_n stand for all capital stocks at the micro level; labor (L) is assumed to be a homogeneous factor of production. The problem of surveying all micro-functions is generally solved by aggregating all micro-capital stocks into an *efficiency capital stock* (Sato 1967, 1975) which yields a macro-production function like

$$(4.2) \qquad Y = g(J, L)$$

where $J = J(K_1, \ldots, K_n)$ represents the efficiency capital stock and is an aggregate of heterogeneous micro stocks. A macro-production function can be derived in the form of equation 4.2, if all marginal productivities of the capital stocks stand in fixed proportions to one another and are hence separable from labor.[2] One can regard an economy as consisting of one firm that has the technology of equation 4.2, or, equivalently, as being comprised of many firms whose *aggregate* technology is represented by the production function. This does not imply that all individual micro-production relations are identical and described by the technical relationship determining equation 4.2 (Pindyck and Rotemberg 1983a, p.1068). The supply side outlined below incorporates such a

1 On the aggregation problem of heterogeneous capital goods see Fisher (1982) and Sato (1975).

2 Compare Section 3.4.2.1. The transcendental logarithmic production function described in Section 3.4.2.2. allows for testing whether inputs are separable from the product mix in gross output, i.e. if the derived demand function for inputs is independent of the output composition. Mohabbat and Desai (1983) and Mohabbat et al. (1984) examine this question for India and South Korea. They include capital, labor, and imports as means of production and find that such a separation is possible which justifies the derivation of input demands from total output; a changing composition of the latter does not alter optimal factor demands.

functional form. Two capital stocks are distinguished from one another: One consisting of domestically produced capital goods while a different one is made up only of imported investment products.

4.2.2. The elasticity of substitution ex post and ex ante

Assume for illustration purposes a production function with a capital stock and labor as arguments. The elasticity of substitution (σ_1) between the two factors, which are rewarded the value of their marginal product, measures the proportionate change in the capital-labor ratio relative to the proportionate change in the wage-rental ratio. Since both of these ratios are assumed to work in the same direction, the value of σ_1 is always positive (unless perfect complementarity is encountered and $\sigma_1 = 0$). Two input factors are characterized to behave like complements in the production process if $0 \leq \sigma_1 < 1$ and they are substitutes if $\sigma_1 \geq 1$. Time-series estimates of the elasticity are likely to *understate* the true possibilities for factor substitution open to new investment, as will be discussed below.

It is important to distinguish the *ex ante* elasticity, which relates to the technological choice of new investment, from the *ex post* elasticity. The ex post elasticity measures the response of factor combinations to relative price changes in previously constructed plants. Further, the elasticity of substitution on the plant level is contrasted to that on the industry or economy level:[3]

(1) *Choice of products ex ante*: Every economy produces more than one good. It can vary the use of inputs by directing new investment so that the composition of output is altered, unless all final products exhibit the same technological production function. If capital is relatively expensive, the product choice falls on commodities that use only scarce capital services and an abundant amount of other factors of production.

(2) *Factor substitution by choice of plant ex ante*: All plants producing one commodity might operate with identical production functions, but it is more likely that different firms employ different technologies. Relative factor prices channel new investment to plants employing a technology that uses the relatively cheap factors of production intensively.

(3) *Factor substitution by choice of technique ex post*: If the product and plant decisions have been made and the capital stock is fixed for a given plant, there might still be the possibility of changing factor proportions ex post. For example, a manufacturing firm operating with fixed equipment and a given operation time can still respond to changing factor prices by modifying the intensity of

---

3 The discussion follows Fitchett (1976) and Winston (1974).

use of the machines. If wage rates decline relative to the user costs of capital, it might be profitable to change the crew operating one machine. While, e.g., one person operated one machine before the price change, it might now pay to have two persons assigned to it (one person working the machine, another one occupied with maintenance and handling of raw inputs and output). An ex post 'instantaneous' substitutability of zero would be present, if one machine had to be operated by a certain amount of personnel and any other crew size were either technically impossible or inefficient.

(4) *Factor substitution by changing the duration of production and idleness ex post*: Finally, a response to factor price changes is possible by changing the duration of operation of the capital stock. Although most often the capital stock itself is included in the formulation of the supply side, the real productive factor is rather the *service flow* from the stock (Dadkhah and Zahedi 1986, Winston 1974). For a given capital stock, the service flows generated from it vary according to the intensity of use (point 3 above) and the time or degree of capacity utilization. For example, if the premium paid to night-shift workers relative to the day-shift wage-rate increases, the cost of operating the machines at night might be higher than the costs of letting the capital stock idle. Consequently, the proportion of time a plant is optimally left idle depends on relative factor prices.

It follows from the above discussion that the requirements for an economy to be characterized by fixed proportions among input factors and total ouput are restrictive. This would only be the case, and a Leontief production function justifiable on the macroeconomic level, if all of the following conditions held simultaneously (Winston 1974, p.158/159):

(a) Sectoral factor use is not affected by the product mix of total output
(b) Factor use is not affected by the choice of techniques because there is only a single kind of plant that can produce each of the commodities.
(c) The intensity of use of each plant is rigid (the crew size fixed).
(d) Utilization of the capital stock is not affected by factor-price changes because capital service flows cannot be varied.

Therefore, it is not necessarily a contradiction to estimate a non-negative elasticity of substitution in production, while an underutilization of the capital stock of an economy is observed. This phenomenon could be encountered if plants have to operate with fixed factor proportions (ex post) and have an optimally idle capital stock with relative factor-price changes, but output can be redistributed (ex ante) between different plants and industries.

The discussion of conditions determining the substitutability between different input factors can be applied to describe the relationship between imported and domestic factors of production. Even if conditions (b) to (d) hold, some substitutability for imported inputs can be achieved by altering the structural composition of the output. The latter is an important determinant of the import intensity of output growth, since the agricultural sector generally requires much less imported inputs than the non-food manufacturing sector (Lopez and Thomas 1988).

We will employ time series analysis to estimate elasticities of substitution between different factor-input combinations in the next chapter. The empirical estimates will therefore reflect both decisions of factor combinations that relate to new investment (ex ante elasticities) and to past investment (ex post elasticities). One can expect that factor input choices are more flexible in regard to the instalment of new machinery than to those already in place. Since estimates of elasticities are a weighted average of these two different input variations, they will "understate the 'pure' ex ante elasticity that is relevant to forecasting price responsiveness in development *planning* because empirical estimates include influence of the lower ex post response of plants already in existence at the time of price change" (Winston 1974, p.160).[4]

A further remark should be made concerning the proposition that the variable in the production function should be rather the *service flow* than the capital stock itself. Since information on capital stock utilization is not available (at least on a more aggregate level), it is most often assumed in empirical studies that the service flow from the capital stock is proportionate to the capital stock itself.[5] Some studies attempt to use a proxy for the utilization of the capital stock; one of the most common is the electricity measure (Foss 1963, Heathfield 1972). Bautista et al. (1981, Chapter 3) report on field studies of capital utilization in selected developing countries. Interestingly, these studies find that the electricity measure as well as other proxies (shift work measure, full capacity at peak method) are all unsatisfactory ways of tracing the actual degree of capacity utilization.

4.2.3. Factors influencing the elasticity of substitution between imported and domestic factors of production

We can readily derive several aspects determining the magnitude of the elasticity of substitution between imported and indigenous means of production from the above discussion. The *ex ante* elasticity of substitution between imported and domestic factors of production will be relatively low, if

4 Winston further concludes that "only if it measured response so long after (one-shot) factor price change that all capital had been replaced since the price change .... would a time series estimate reflect the full opportunity for factor substitution" (Winston 1974, p.160).

5 Compare Dadhkah and Zahedi (1986, fn 1).

- domestic factors of production are immobile both within an industry (at the plant level) and between different sectors of the economy. This reduces the scope of adjustment through shifting production to those plants and sectors that are relatively independent of imported inputs;
- production of previously imported factors of production is possible, but only at considerably higher costs of production; this could be the case for example if the developing country lacks the technological capability to substitute for imported intermediate and capital goods;
- specific non-traded factors of production are not available (sites of natural resources).

The *ex post* elasticity of substitution is low, if both the intensity and the utilization of existing factors of production can only be varied marginally due to technological or institutional (e.g. strict regulation of working hours) determinants.

Substitution possibilities for imported capital and intermediate goods are determined by factors which are both endogenous as well as exogenous to the developing economy. The natural resource endowment of a country is obviously an exogenous factor, while the trade and development strategy, for example, are policy instruments chosen by each country individually. Inward oriented development strategies are generally geared towards reducing the importance of imported products in the total consumption of different goods. These policies aim at decreasing the reliance on imports. We will explore the question whether such a strategy was indeed successful in raising the elasticities of substitution or not in the empirical part of this study.

Some remarks should be made regarding the *constancy* of the elasticity of substitution. Returning to the distinction between ex post and ex ante elasticities, the concept is difficult to defend if the (macro) ex-ante elasticities are looked at. These elasticities build on redistribution decisions within an economy. It is obvious that consecutive relative price changes of the same magnitude may cause different relocation decisions, depending on the stage of specialization itself. Hence, if an economy adjusts successfully to reduced import levels by shifting production to less import-intensive sectors (e.g. the agricultural sector), it cannot be taken for granted that a further reduction of imports will be adjusted to in the same way.

Additionally, both the ex ante and the ex post elasticities are influenced by policy decisions, which will alter them over time. The constancy assumption of the elasticity of substitution which we adopt is therefore rather operational. Estimated coefficients can be viewed as representing 'average' elasticities over a certain period of time. Nevertheless, we will make an attempt to measure whether significant changes of the parameter took place in the period under consideration.

4.3. The nested production function

4.3.1. Disaggregation of imports

In his theoretical analysis, Van Bochove (1982) includes *total* imports as a factor of production in the domestic output function. He chooses a homogeneous function of the form (Van Bochove 1982, pp.67-68)

$$(4.3) \qquad Q = f(V,M)$$

and

$$(4.4) \qquad V = g(K,L),$$

with V as domestic value added and M, K, and L standing for imports, the capital stock, and labor. It is thus assumed that capital and labor are separable from total imports. A separability of domestic factors of production and imports could be deemed appropriate, if total imports are a homogeneous commodity. Obviously, complete homogeneity (i.e. a country imports only one commodity) cannot be required, since it will never be achieved in reality. A necessary condition for the treatment of imports as a homogeneous good would be, though, if they were employed in the same way in the importing country. For example, an import-list consisting of 90% of intermediate goods could justify the assumption.

We observed that a remarkable share of traded goods, especially imports of developing countries, is conducted in capital products. These goods are invested and supplement those domestically produced in investment projects. Estimating the capital stock by means of an aggregation process which employs figures on total gross domestic investment and depreciation rates produces an aggregate measure of the total capital stock. Postulating that total imports are a factor of production in addition to a derived capital stock, as outlined above, results in biased estimates per se (independently of whether the separability assumption is true or not), since imported capital goods are counted twice.[6] Considering the importance of capital goods in the total import bill, a separability of aggregate imports from domestic means of production appears to be

6 As described in Chapter 3, Bergsten et al. (1985) estimate a Cobb-Douglas production function incorporating the fixed capital stock, the active labor force and current and lagged imports. They approximate the capital stock by accumulating five years of gross fixed capital formation, including the 'current' year t. Their results are therefore biased. One could avoid this problem by assuming the current operating capital stock K_t to be aggregate net investment up to the period (t-1) (e.g. Dadkhah and Zahedi 1986, Park 1985).

unjustified, as then short run fixed (domestic and imported capital goods) and short run variable (labor and imported inputs) factors are not distinguished from each other.[7]

For this reason, we disaggregate real imports and examine intermediate and capital goods separately. Two capital stocks are distinguished: K_d consists of investment undertaken at home and K_f represents accumulated (and depreciated) imports of capital goods.[8]

4.3.2. A three-level nested CES function

Following Michalopoulos (1975), Marquez (1984, 1985), and Taylor (1979), we assume that the two capital stocks (K_d and K_f) are separable from other input factors. They form an efficiency capital stock on the first production level. The combination of the efficiency capital stock with the factor labor gives the second stage of the production function, which is by definition equal to domestic value added. Finally, we introduce imported intermediate imports at the third stage, in which gross output is modelled as a function of domestic value added and imported intermediate products.[9] This third stage is the hyper function. A three-level production function of the following form is subject to analysis:[10]

The efficiency capital stock J_t is defined as

$$(4.5) \qquad J_t = J(K_d, K_f),$$

which joins together with labor (L) to generate domestic value added (Y),

$$(4.6) \qquad Y_t = Y[L, J(K_d, K_f)].$$

---

7 As mentioned in Section 3.4.2.3., several authors (Burgress 1974a and 1974b, Mohabbat and Desai 1983, Mohabbat et al. 1984) reject the separability of aggregate imports as an input from the indigenous capital stock and labor.

8 Farzin (1988) and McKinnon(1964) distinguish between the two different capital stocks, assuming a fixed-proportions production function. Marquez (1984, 1985), Michalopoulos (1975), and Taylor (1979) employ them in a nested CES version. Taylor does not estimate the function, though, and Marquez derives his estimating equation under the assumption that σ_k is small. It should be noted that the capital stocks are included in the production function rather than the service flows from the former. Since most proxies of the service flow have proven unsatisfactory (compare Bautista et al. 1981, Chapter 3), we assume that the former is proportionate to the capital stock.

9 Compare Bruno and Sachs (1985, p.57), Conway (1986, p.283), Findlay and Rodriguez (1977, p.209), and König (1990, p.86).

10 Obviously, home-made intermediate inputs enter production as well; it is assumed that they are produced using K_d, K_f, L, and M_{Int}, so that they are subsumed in the hyper function (compare Conway 1986, p.239, fn 4).

Total gross output, Q_t, then becomes

$$(4.7) \qquad Q_t = Q\left[Y\{L, J(K_d, K_f)\}, M_{Int}\right].$$

M_{Int} stands for imports of intermediate goods. Assuming CES aggregator functions over Q(.), Y(.), and J(.), and the hyper function to be linearly homogeneous of degree 1 gives[11]

$$(4.8) \qquad Q_t = \left[c\, Y_t^{-\tau} + (1-c) \left[e^{z_M t} M_{Int}\right]^{-\tau} \right]^{-(1/\tau)}$$

$$(4.9) \qquad Y_t = \left[b\, J_t^{-\Phi} + (1-b) \left[e^{z_L t} L_t\right]^{-\Phi} \right]^{-(1/\Phi)}$$

$$(4.10) \qquad J_t = \left[a \left[e^{z_{Kd} t} K_d\right]^{-\theta} + (1-a) \left[e^{z_{Kf} t} K_f\right]^{-\theta} \right]^{-(1/\theta)}$$

where

a,*b*, and *c*	are distribution parameters;	
z_{Kd}, z_{Kf}, z_L, and z_M	represent disembodied technological progress for each of the factor inputs;[12]	
θ, Φ, and τ	are substitution parameters with:	
	$\sigma_k = 1/(1+\theta)$	the elasticity of substitution between the two capital stocks;
	$\sigma_l = 1/(1+\Phi)$	the elasticity of substitution between the efficiency capital stock and labor;
	$\sigma_m = 1/(1+\tau)$	the elasticity of substitution between domestic value added and intermediate imports.

---

11 Theoretically, only the hyper function can be analyzed with respect to the degree of homogeneity in a multi-level functional system. There always exists a linear homogeneous function in J(.) and Y(.), so that the hyper function Q(.) is homogeneous of degree μ. See Van Bochove (1982, p.68/69) for a proof. This implies that the functional specification (and estimation) of equations 4.5 and 4.6 are independent of the linear homogeneity restriction.

12 Alternative to specifying factor-specific technical progress, one could model Hicks neutral technological progress (e.g., Conway 1986, p.239) or include an efficiency parameter in the CES formulation (compare, e.g., Marquez and Pauly 1987, p.94).

The production decision is disjoined in three stages. The two parameters measuring the importance of imported factors of production are σ_k and σ_m in equations 4.8 and 4.10. The concept of separability validates the application of optimization conditions to each input decision level, *independently* of the others. It is therefore possible to leave estimation and determination of σ_l (the elasticity of substitution between labor and the efficiency capital stock) aside.

4.3.3. The optimal composition of the efficiency capital stock

Since the production decisions are separable from each other, we consider equation 4.10 first, where the efficiency capital stock is in a functional relationship of two stocks. One consists of imported investment goods, the other is made up of domestically manufactured ones. We assume that firms minimize costs of production subject to a given efficiency capital stock.[13] The Lagrangian can then be expressed as

$$(4.11)\quad L_J = P_{Kd}K_d + P_{Kf}K_f + \Omega_J\left\{\bar{J}_t - \left[a\left[e^{z_{Kd}t}K_d\right]^{-\theta} + (1-a)\left[e^{z_{Kf}t}K_f\right]^{-\theta}\right]^{-1/\theta}\right\},$$

where P_{Kd} and P_{Kf} represent the user costs of capital and $\bar{J}_t$ stands for the actually realized efficiency capital stock.[14] $\bar{J}_t$ is the production constraint and already determined for any given firm. The derivation of the Lagrangian with respect to K_d, K_f, and Ω_J describes the first order conditions for a cost minimum:

$$(4.12)\quad \frac{\partial L_J}{\partial K_d} = P_{Kd} - \Omega_J\frac{\partial J}{\partial K_d} = 0$$

$$(4.13)\quad \frac{\partial L_J}{\partial K_f} = P_{Kf} - \Omega_J\frac{\partial J}{\partial K_f} = 0$$

$$(4.14)\quad \frac{\partial L_J}{\partial \Omega_J} = \bar{J}_t - \left[a\left[e^{z_{Kd}t}K_d\right]^{-\theta} + (1-a)\left[e^{z_{Kf}t}K_f\right]^{-\theta}\right]^{-1/\theta} = 0\ .$$

---

13 The usual argument when applying a cost-minimzation approach to macro-production functions rests with the aggregation of all micro-functions of the individual firms which face given factor prices.

14 Time subscipts (of period t) are omitted for convenience, if the notation becomes too confusing. Subscripts of other than period t are specified.

Combining 4.12 and 4.13 yields

$$(4.15) \qquad \frac{P_{Kd}}{P_{Kf}} = \left[\frac{a}{1-a}\right]\left[\frac{K_d^*}{K_f^*}\right]^{(-1-\theta)} e^{(z_{Kf}-z_{Kd})t\theta},$$

where an asterisk denotes a 'desired' or 'optimal' quantity. Rearranging equation 4.15 gives

$$(4.16) \qquad \frac{K_d^*}{K_f^*} = \left[\frac{a}{1-a}\right]^{\sigma_k}\left[\frac{P_{Kf}}{P_{Kd}}\right]^{\sigma_k} e^{(z_{Kf}-z_{Kd})t\sigma_K\theta}.$$

We include a technological progress term for each of the two capital stocks in the production function for the efficiency capital stock (4.10), since the assumption of Hicks-neutral technological change is restrictive.[15] Investment goods which enter into the capital stock consisting of imported capital goods might differ significantly from those that form the domestic capital stock. It is thus appropriate to allow the rates of technological progress to differ from each other (z_{Kd} and z_{Kf}); if they are equal, the technological progress term in equation 4.16 cancels out.

Equation 4.16 describes the ratio of the two desired capital stocks as a function of the relative price (each representing user costs) and the technological progress term.[16] But observed capital stocks are unlikely to be desired capital stocks, since the ex post substitution of capital goods is limited (they cannot simply be sold again). A downward adjustment of any one capital stock is mostly driven by depreciation. Additionally, the time lags due to order and delivery support the hypothesis that observed capital stocks are not desired long-run equilibrium capital stocks. We employ a partial adjustment process of the following form:

---

15 Compare Sato (1967) and Nerlove (1967) for modelling non-neutral technical progress in CES production functions.

16 It is important to note that the efficiency capital stock J(.) is not equal to a measure of the total capital stock K_t ($K_t = K_d + K_f$). Since imported and indigenous factor inputs are not assumed to be perfect substitutes a priori, they have different weights in forming the composite efficiency capital stock. If J(.) is defined to be K_t, an identity would be estimated. Some authors employing the nested CES function do not distinguish explicitly between the aggregate measure K_t and an efficiency measure J(.). A similar problem arises when one considers the composition of the change of the stock, i.e. net investments. Marquez (1984, pp.60-62), for example, states that he assumes domestic and foreign capital goods to be imperfect substitutes, but he uses figures for net investment (including both the domestic and the imported capital goods) in his estimating equations.

$$(4.17)\qquad \frac{K_d}{K_f} = \left[\frac{K_d^*}{K_f^*}\right]^{\lambda_1} \left[\frac{K_{d,t-1}}{K_{f,t-1}}\right]^{(1-\lambda_1)},$$

where λ_1 stands for the speed of adjustment with $0<\lambda_1\leq 1$, with $\lambda_1=1$ for immediate adjustment to the desired factor input ratio. Combining equations 4.16 and 4.17 gives:

$$(4.18)\qquad \frac{K_d}{K_f} = \left[\frac{a}{1-a}\right]^{\sigma_k\lambda_1} \left[\frac{P_{Kf}}{P_{Kd}}\right]^{\sigma_k\lambda_1} e^{(z_{Kf}-z_{Kd})t\lambda_1\sigma_K\theta} \left[\frac{K_{d,t-1}}{K_{f,t-1}}\right]^{(1-\lambda_1)}.$$

We will estimate this equation in the next chapter. Except for the rates of technical progress (z_{Kd} and z_{Kf}), all parameters included in equation 4.10 can be identified. The long run elasticity of factor substitution is defined as σ_k, while the short run elasticity is $\lambda_1\sigma_k$. The estimates of the structural parameters could be used to derive both the efficiency capital stock[17] as well as the user cost thereof, applying the duality property of the CES production function.[18]

4.3.4. Optimal production of gross output

The derivation of the equation that enables us to estimate σ_m (the elasticity of substitution between domestic value added and imported intermediate inputs) proceeds analogously to the cost minimization procedure applied above. The Lagrangian of equation 4.8 becomes

---

17 A transformed estimate of J_t can be defined as

$$\hat{J}'_t = \left\{ a\, K_d^{-\theta} + (1-a) \left[e^{(z_{Kf}-z_{Kd})t}\, K_f\right]^{-\theta} \right\}^{-(1/\theta)}$$

with

$$\hat{J}_t = \hat{J}'_t\, e^{z_{Kd}t}$$

as can be easily derived from equation 4.10. This specification could be used in the next step of estimation (first order conditions of equation 4.9).

18 For a derivation of the cost function from a CES production function compare Appendix B.

$$(4.19)\quad L_Q = P_Y Y + P_{MInt} M_{Int} + \Omega_Q \left\{ \bar{Q}_t - \left[c\, Y_t^{-\tau} + (1-c)\left[e^{z_M t} M_{Int}\right]^{-\tau} \right]^{-1/\tau} \right\}.$$

For cost minimizing firms the first order conditions are described by

$$(4.20)\quad \frac{\partial L_Q}{\partial M_{Int}} = P_{MInt} - \Omega_Q \frac{\partial Q}{\partial M_{Int}} = 0$$

$$(4.21)\quad \frac{\partial L_Q}{\partial Y} = P_Y - \Omega_Q \frac{\partial Q}{\partial Y} = 0$$

$$(4.22)\quad \frac{\partial L_Q}{\partial \Omega_Q} = \bar{Q} - \left[c\, Y_t^{-\tau} + (1-c)\left[e^{z_M t} M_{Int}\right]^{-\tau} \right]^{-1/\tau} = 0.$$

Combining equations 4.20 and 4.21 results in

$$(4.23)\quad \frac{P_Y}{P_{MInt}} = \left[\frac{c}{1-c}\right] \left[\frac{Y^*}{M_{Int}^*}\right]^{(-1-\tau)} e^{z_M t \tau}.$$

Rearranging terms and assuming a partial adjustment process as depicted when considering the ratio of capital stocks (equation 4.17) produces

$$(4.24)\quad \frac{Y}{M_{Int}} = \left[\frac{c}{1-c}\right]^{\sigma_m \lambda_2} \left[\frac{P_{MInt}}{P_Y}\right]^{\sigma_m \lambda_2} e^{z_M t \lambda_2 \sigma_m \tau} \left[\frac{Y_{t-1}}{M_{Int,t-1}}\right]^{(1-\lambda_2)},$$

with λ_2 as the speed at which the ratio on the left hand side of the equation adjusts to its desired level if a change in the relative price term occurs ($0 < \lambda_2 \leq 1$).

The basic reduced form equations are represented by equations 4.18 and 4.24 above. They will be estimated for individual developing countries in the next chapter.

4.4. Summary

Two reduced form equations suitable for parameter estimation were derived in this chapter. They were deduced from a three-level constant elasticity of substitution function under the assumption that firms minimize production costs subject to an output constraint. Two different capital stocks, labor, and intermediate inputs are included in the supply-side specification. The formulation is flexible, as some input pairs can show a complementarity relationship while others can be substitutes for each other. Even if perfect complementarity between one input pair exists, this does not imply overall factor inflexibility, which is encountered in trade-gap models.

5. Estimation of elasticities of substitution between domestic and imported means of production

5.1. Introductory remarks

After discussing methodological problems associated with estimating import demand functions and deriving the estimating equations, we now turn to the empirical analysis. Some of the endogenous and exogenous variables included cannot be observed directly and proxies have to be used instead. For this reason, we deal with some issues of data generation and assumptions about them explicitly in the text instead of deferring them to the appendix.

Section 2 of this chapter is concerned with measurement problems of the internal price of imported products. The two different approaches applied alternatively in the estimations are described. One introduces a protection proxy which is calculated from available statistics. The other models the impact of tariff and non-tariff barriers on imports indirectly as a function of foreign exchange availability. In Section 3, we report estimations of the production function describing the efficiency capital stock as a function of the imported and the domestically produced capital stock. The sensitivity of the estimation results with respect to data generation, depreciation rate, and the user cost definition is examined. Emphasis in the discussion of the calculated parameters lies on the elasticity of substitution between the two capital stocks. Section 4 then turns to estimates of the hyper function, in which domestic value added and intermediate imports combine to produce the gross output of the economy. The chapter concludes with a condensation and interpretation of the results obtained.

5.2. Measurement of the domestic price of imported goods

5.2.1. The domestic price of imported goods

Equations 4.18 and 4.24, which are the basis for the estimations in this chapter, both include a price ratio of a domestic mean of production relative to an imported one. While the domestic price of home-made capital goods and of value added can be gathered from available statistics, it is much more difficult to assess the domestic price of imported capital and intermediate goods as the effect of protectionist policies has to be included. It is possible to calculate the border price of imported capital and intermediate goods by aggregating and deflating country specific trade data, but the effect of protectionist policies cannot be observed directly for such broad bundles of import groups.

We discussed in Chapter 3 the fact that calculations of the effective rate of protection based on a price comparison of the domestic price of capital and intermediate goods and world market prices of their imported counterparts (taking into account subsidies and tariffs on imported inputs) cannot be used since they are based on the perfect homogeneity assumption and assume that the law of one price holds.[1] As argued in Chapter 2, it is unlikely that perfect substitutes are manufactured in developing countries for the bundle of imported capital and intermediate goods. While it does make sense to compute the effective rates of protection for single goods where the homogeneity assumption can be regarded as fulfilled,[2] it is less convincing to apply the price comparison to the broad product categories investigated here.

If imported and domestically produced goods are not perfect substitutes, their relative price can change over time. The time path of the relative price between imperfect substitutes depends on the degree of substitutability, changes in protective measures or the exchange rate, and shifts of demand and supply functions of both the imported and the domestic goods.[3]

As argued in Chapter 3, the effect of protection can be modelled using two different approaches. One uses a protection proxy, the other specifies a trade policy reaction function of the government. We apply both of these concepts alternatively in the estimations.

5.2.2. Derivation of a protection proxy

The protection proxy uses information on the structure of protection and proceeds from import charges to construct a proxy for the price-effect of protectionist policies. Nominal tariff and total charges data are collected by UNCTAD in the Tariff Information System (TIS).[4] Un-

---

1 Compare the studies cited in the survey by Goldstein and Khan (1985, p.1045) which test and reject the validity of the law of one price for both aggregated and disaggregated commodity groups.

2 See, for example, Corsepius and Schipke (1989), who compare domestic prices of micro-computers manufactured in Brazil to the world prices of micro-computers.

3 See, e.g., Shiells et al. (1986, p.497).

4 *Tariffs* are considered by UNCTAD to consist of customs duties and include fiscal duties when such exist. *Total charges* comprise tariffs and additional charges, such as customs surcharges and surtaxes, stamp taxes, certain other fiscal charges, and taxes on foreign exchange transactions. Compare UNCTAD (1987b and 1987c) for country specific infor-

fortunately, data are only available for all countries in 1987/88 and for some since 1985. Table 5.1 compares data on total charge rates levied on imports of primary goods and manufactures for those countries in the sample for which data could be obtained from the TIS in 1985 and 1987/88. The column entitled 'ratio' is the total charge rate on primary goods divided by the total charge rate on manufactures.

One can gauge from the table that the structure of the charge rates remained relatively stable from 1985 to 1987/88, even in those cases where a major revision of the level of nominal charges occurred (Argentina, Chile, Columbia, Mexico). The simplifying assumption is made, on that account, that the *structure* of nominal (weighted)[5] protection from import charges observed in 1987/88 applied to the whole period considered (1972 to 1987). This does not imply, however, that the protectionist effect on each import category itself remained constant. Rather, we assume that tariff and non-tariff barriers were raised and lowered on all import categories alike, so that the relation between the protectionist effects on the different import categories remained stable.

It is then possible to obtain estimates of the *implicit average effective import taxes* using data on tax revenues from imports which are supplied by the IMF (1990b).[6] The implicit tax rate is defined as the ratio of the government revenue from imports (measured in current prices and at current exchange rates) and the nominal import bill (UNCTAD 1989, p.96). Application of the structure of nominal charges to data on nominal imports of capital, intermediate, and consumer goods as well as total revenues from import charges, enables us to compute the implicit average effective tax rates for each import category separately.[7]

---

mation on the extent of coverage of additional charges. Since the total charge rates also include some non-tariff measures, these figures are chosen to compute the structure of protection.

5 We use nominal *weighted* protection data in order to account for the influence of a changing composition of trade on calculations of the nominal rate of protection for broad product groups: Since a bundle of many products is looked at, the simple mean average nominal charge of the Customs Co-operation Council Nomenclature (CCCN) weighted with imports of the reporting country would underestimate the leverage of import charges, since imports facing high import charges attain a low weight due to low import volumes. Or put the other way, average nominal charges on imports would change not only due to a change in the tariff line but also due to a changing composition of trade. In order to minimize this trade composition effect, a weighted protection measure is employed, where each nominal charge on the CCCN line is weighted with the import structure of *all* developing countries.

6 The IMF (1990b) reports revenues on total import duties (including customs duties) and other import charges.

7 It is assumed that duty exemptions and the tax collection effort are independent of the type of the imported good.

Table 5.1:
Structural change of nominal charge rates, 1985-1987/88

Country	Year	Total charge on primary goods[1] (per cent)	Total charge on manufactured goods[1] (per cent)	Ratio[2]
Argentina	1985	24.7	29.2	.85
	1988	35.0	41.6	.84
Chile	1985	35.6	36.8	.97
	1988	20.9	21.8	.96
Columbia	1985	71.8	87.4	.82
	1988	43.5	50.2	.87
Indonesia	1985	15.2	21.8	.70
	1987	14.7	19.6	.75
Malaysia	1985	8.8	16.5	.53
	1988	7.7	16.6	.46
Mexico	1985	24.9	37.1	.67
	1988	13.4	17.1	.78
Peru	1985	44.0	57.3	.77
	1987	53.2	71.5	.74
Philippines	1985	36.9	38.5	.96
	1987	31.9	33.5	.95
Uruguay	1985	29.9	33.2	.90
	1988	26.6	29.2	.91
Venezuela	1985	31.6	29.4	1.07
	1988	34.5	32.4	1.06

Source: UNCTAD Tariff Information System.

1 Mean total charge (not weighted by the composition of trade) for all tariff lines defined as corresponding to primary and manufactured goods. Compare UNCTAD (1987b and 1987c).

2 Ratio of total charge on primary goods to total charge on manufactured goods.

Table 5.2: Comparison of nominal charge rates and implicit average effective tax rates on imports (per cent)

Country	Rate	Capital goods	Interme-diate goods	Consumer goods
Argentina	nominal total charge[1] (Feb. 1988)	43.0	36.5	39.6
	implicit average tax[2]	20.0	17.0	18.4
Brazil	nominal total charge (Nov. 1986)	67.8	71.5	93.4
	implicit average tax	7.0	7.4	9.6
Chile	nominal total charge (March 1988)	15.5	20.6	24.2
	implicit average tax	11.3	15.1	17.6
Columbia	nominal total charge (Sept. 1986)	69.5	67.3	89.2
	implicit average tax	18.7	18.1	24.0
Ecuador	nominal total charge (Dec. 1986)	30.4	30.7	67.6
	implicit average tax	13.2	13.3	29.3
Indonesia	nominal total charge (Sept. 1987)	18.0	12.9	29.3
	implicit average tax	4.9	3.5	7.9
Malaysia	nominal total charge (June 1987)	10.2	11.1	19.2
	implicit average tax	4.5	4.9	8.4
Mexico	nominal total charge (May 1988)	17.3	10.5	19.2
	implicit average tax	11.7	7.1	13.0
Peru	nominal total charge (Sept. 1987)	57.2	50.0	74.8
	implicit average tax	30.6	26.7	40.0
Philippines	nominal total charge (Oct. 1987)	29.1	26.8	42.4
	implicit average tax	18.1	16.7	26.3
Uruguay	nominal total charge (Feb. 1988)	24.2	28.1	31.7
	implicit average tax	14.6	17.0	19.1
Venezuela	nominal total charge (March 1988)	24.3	29.0	48.5
	implicit average tax	16.2	19.3	32.3

Source: Own calculations based on UNCTAD Tariff Information System, UN Comtrade database, IMF (1990b).

1 The 'nominal total charge' is the average charge rate of all SITC lines defined as capital, intermediate, and consumption goods as defined in Appendix C, Table C.2. Each total charge rate is weighted with the import composition of *all* developing countries.

2 The 'implicit average tax' is calculated by applying the structure of the nominal total charge-rates to data on import duty collection (valued at current prices and the current exchange rate) and nominal imports of the three import categories. It is assumed that the collection effort of taxes is equally spread across all import categories.

Table 5.2 compares the nominal average charge rate to the implicit average effective tax rate on imports in 1987. Except for Argentina, imported consumer goods face the highest level of charge rates, ranging up to twice that of intermediate goods for Ecuador, Indonesia, Malaysia, and Mexico. The difference between the nominal total charge levied on intermediate goods and capital goods tends to be quite low. The implicit tax rate on the individual import categories are much lower than the nominal rates. Probably, this is due to low tax collection efforts or tax exemptions (Amelung and Sell 1991).[8] The difference between nominal total charge rates and the implicit average tax rate is most striking for Columbia and Brazil, where they diverge by a factor of four and ten.

Although several non-tariff measures are included in the implicit average effective import tax figure (stamp tax, import surcharges, foreign exchange taxes etc.), the effect of quantitative restrictions is not registered. Nevertheless, one might assume that quantitative restrictions on imported capital and intermediate goods are strengthened and weakened concurrently with changes in the average effective import tax (Conway 1986, p.240). This premise can be justified on empirical grounds. UNCTAD (1989, pp.122-124) examines the change in total import charges and the change in the use of quantitative restrictions (frequency of application by tariff line) between 1985 and 1988. For the countries considered here, UNCTAD calculations are reproduced in Table 5.3 (Indonesian data are missing). A concurrent movement of the charge rate and quantitative restrictions cannot be observed for only two countries. Malaysia raised the application of quantitative restrictions marginally between 1985 and 1988, while the nominal charge rate remained constant. The charge rate was lowered slightly in Uruguay, whereas quantitative controls remained stable. In the other nine cases quantitative restrictions were raised and lowered jointly with the modification of the charge rate.[9]

8 The discrepancy between nominal charge rates and implicit effective rates suggests positively redundant tariffs. Tariffs are positively redundant if the combined effect of all tariff and non-tariff barriers produces an effective protection rate that is lower than the nominal legal tariff. Although the implicit average tax rates and not the effective rates of protection are computed here (which would be necessary to calculate true tariff redundancies) the results are in line with studies that have found positive redundancies (Amelung and Sell 1991, Langhammer 1988).

9 Whalley (1989, pp.7-9) reports that tariff cuts in the second half of the 1980s went hand in hand with the reduction of quantitative controls in Argentina, Brazil, Mexico, and the Philippines.

Table 5.3:
Comparison of import charge rates and leverage of quantitative restrictions, 1985-1988

Country	Import charges[1]		Quantitative restrictions[2]	
	1985	1988	1985	1988
Argentina	28	26	50	31
Brazil	81	42	34	16
Chile	35	17	1	.2
Columbia	83	48	96	74
Ecuador	50	49	38	27
Malaysia	14	14	4	5
Mexico	34	16	19	12
Peru	54	66	50	53
Philippines	38	33	53	43
Uruguay	32	29	1	1
Venezuela	30	33	17	18

Source: UNCTAD (1989, pp.122-124).

1 Tariffs and equivalent import charges (unweighted averages for all products) as a percentage of the value of imports.

2 Discretionary licensing and import prohibitions (unweighted frequency of application by tariff line in per cent).

We include the protection proxy in order to calculate the internal price of imported goods, which is needed in the estimations. This calculation is done in the following way: The internal price of imported goods (P_i) is equal to the product of the border price (P_i') and the combined effect of all tariff and non-tariff barriers:

$$(5.1) \qquad P_i = P_i' \, (1+t_i).$$

The calculated implicit average effective tax rate on the import categories is used as a proxy for t_i in equation 5.1.

5.2.3. Foreign exchange availability and the internal price of imported goods

The protection proxy computed to reflect the magnitude and influence of tariff and non-tariff measures in order to reflect the true internal price of imported goods is, however, based on several strong assumptions. They are (a) that the structure of nominal total charge rates also reflects the structure of the protection effect if all tariff and non-tariff barriers are taken into account, (b) that the structure of protection stayed relatively constant over the period under consideration, and (c) that the variance of the implicit effective tax rate also reflects the price effect of quantitative import controls on imported goods.

Quite a few countries in the sample underwent shorter or longer liberalization attempts during the 1970s and 1980s (compare Table 2.7 in Chapter 2). One of the first measures in most liberalization attempts is the gradual *replacement* of quantitative restrictions by tariffs. Although empirical support can be traced for the fact that charge rates and quantitative restrictions varied jointly from 1985 to 1988, such a general statement is nevertheless open to criticism. Alternatively to including the protection proxy, we model the combined effect of tariff and non-tariff barriers indirectly. The sensitivity of the computations with respect to the protection proxy can then be tested.

One of the most important variables determining government trade policies is the availability of foreign exchange to finance imports. Faini et al. (1988, p.10) state that "whereas the pervasive presence of quantitative restrictions in developing countries is often motivated by the desire to foster domestic industrialization, the considerable variability of quantitative restrictions can be attributed mostly to balance of payments considerations".

The balance of payments difficulties encountered by many developing countries in the 1980s were also primarily confronted by trade-restricting measures. Laird and Nogués (1988a, 1988b) note that all highly indebted countries responded to the payments crisis by increasing import barriers as a way of generating a foreign exchange surplus.[10] They find that the increase in import barriers was affected mainly by introducing generalized mechanisms of import licensing (Laird and Nogués 1988b, p.2). The policy reversal also applied to countries - including Argentina, Chile, Columbia, Peru, and the Philippines - that had introduced trade liberalization policies prior to the outbreak of the debt crisis.

10 See also Williamson (1990).

We assume here that governments set the domestic distortion $(1+t_i)$ according to a reaction function, which is dependent on foreign exchange availability:[11]

$$(5.2) \qquad (1+t_i) = m_0 \; FX^{m_1},$$

where FX is a variable measuring foreign exchange availability.[12] The constant m_0 incorporates the price effect of trade barriers invariant to the foreign exchange variable and primarily determined by basic policy considerations and the development strategy. The parameter m_1 can be termed the 'foreign exchange elasticity of the rate of protection'. It is supposed to be negative - a fall in FX causes a rise in the rate of protection. We incorporate function 5.2 alternatively to the protection proxy in the estimations below.

5.3. Imported and domestically produced capital goods

5.3.1. The stochastic equation

Equation 4.18 contains the first-order conditions for cost minimizing firms with respect to the input choice between the capital stock made from domestic production and that consisting of imported capital goods.

A multiplicative error of the exponential form in equation 4.18 is assumed. Taking logarithms yields the following stochastic equation:

$$(5.3) \qquad \ln\left[\frac{K_d}{K_f}\right] = a_0 + a_1 \ln\left[\frac{P_{Kf}}{P_{Kd}}\right] + a_2 t + a_3 \ln\left[\frac{K_{d,t-1}}{K_{f,t-1}}\right] + u_3$$

11 We discussed in Chapter 3 the other possibility of modelling a government reaction function, which incorporates the influence of the foreign exchange availability on the speed of adjustment of import volumes. Compare also fn 41 in this chapter.

12 Equation 5.2 is only an approximation of the true structural relationship. Theoretically, $(1+t_i)$ should have a lower bound of 1 if foreign exchange receipts are abundant and it should be high in the reverse case. But the approximation can be defended since foreign exchange receipts neither become infinite nor are they zero. Compare Moran (1989, pp.283-284) for a similar formulation.

with

$$a_0 = \sigma_k \lambda_1 \ \ln\left[\frac{a}{1-a}\right]$$

$$a_1 = \sigma_k \lambda_1$$

$$a_2 = (z_{Kf} - z_{Kd}) \lambda_1 \sigma_k \theta$$

$$a_3 = (1-\lambda_1).$$

K_d and K_f again represent the two capital stocks, t is the time index, while P_{Kd} and P_{Kf} stand for their user costs. The parameter a_1 is a combination of the long run elasticity of substitution (σ_k) between the two capital stocks and the adjustment parameter (λ_1) and represents the estimate of the short run elasticity of substitution. Tests on the magnitude of this parameter concentrate on whether it is significantly different from zero, which would indicate that short run substitution possibilities exist in the respective developing country, and whether it is significantly different from 1. If a value for the short run elasticity is found significantly larger than zero but also significantly smaller than 1, the two products are complements in the short run. Tests on σ_k will also focus on this distinction. If the long run elasticity is not significantly different from zero, this would imply that the technical relationship between the two capital stocks can be described by a fixed-proportions production function.[13] If, on the other hand, the long run estimate of σ_k turns out to be significantly different from zero but not from 1, function 4.10 can be approximated by a Cobb-Douglas function for the developing economy. Since annual data are used to estimate equation 5.3, the mean adjustment time of the capital stock ratio to its desired level becomes $a_3/(1-a_3)$ years. Except for the rates of technical progress (z_{Kd} and z_{Kf}), all parameters included in equation 4.10 can be estimated and identified with equation 5.3.

In the computation of both endogenous and exogenous variables included in equation 5.3, a variety of assumptions have to be made. We examine the sensitivity of the estimated coefficients to these assumptions by running separate regressions which alter postulated parameters and data sources. The specifications of the included variables are discussed in detail below, but it might be helpful to describe briefly the different modifications which are examined consecutively (see Table 5.4) . In the *reference* estimation, industrial production statistics are used to derive estimates of the capital stock made up of domestically produced machinery and equipment, postulating a

13 Lucas (1989, p.145) estimates CES production functions for Indian industries and notes that "in contexts where the implied elasticity of substitution .... proves indistinguishable from zero, a fixed coefficient approach is adopted".

Table 5.4:
Specification of the reference estimations and modifications (capital goods)

Modification	Depreciation rate	User cost concept	Investment data source	Incorporation of the effect of protection
Reference	12.5%	replacement cost	industrial statistics (production)	protection proxy
Modification 1	10.0%	replacement cost	industrial statistics (production)	protection proxy
Modification 2	12.5%	alternative (including interest and depre-ciation rate, capital gains)	industrial statistics (production)	protection proxy
Modification 3	12.5%	replacement cost	national income sta-tistics (expenditure)	protection proxy
Modification 4	12.5%	replacement cost	industrial statistics (production)	foreign exchange availability

depreciation rate of 12.5%. The price of the two capital stocks is chosen to be the replacement cost and the protection proxy described above is incorporated to reflect the price effect of tariff and non-tariff barriers. In the *first modification*, the postulated depreciation rate is altered to 10%, while all other data aspects remain as in the reference case. The *second modification* is concerned with an alternative user cost concept of the capital stocks, while the *third modification* employs expenditure data on investment instead of production statistics. Finally, the *fourth modification* incorporates a trade policy reaction function of the government instead of including the protection proxy.

5.3.2. Reference estimations

5.3.2.1. Data and parameter assumptions

The ratio of the domestically produced capital stock to that made of imported ones is the endogenous variable in equation 5.3. Data on actual capital stocks are not available, so that researchers most often assume a beginning year capital-output ratio and construct time-series for the stocks.[14] A different approach has to be used here, as we are dealing with two different stocks. We construct time series of capital stocks by accumulating net investment per year (gross investment minus depreciation).[15] It is assumed that every investment depreciates at a constant rate. A depreciation rate of 12.5% is applied in the reference scenario, with no depreciation taking place in the period of purchase. It is assumed that the capital goods are installed in the year of production and purchase, so that inventory changes are neglected.[16]

In the reference case, real domestic production data of capital goods (net of exports) in 1980 dollars are used to deduce figures for the domestically produced capital stock K_d. Real output of the manufacturing industries 381 to 385 of the International Standard Industrial Classification net of exports are employed as a proxy for hardware installation. Correspondingly, imports of capital goods which form K_f are defined as matching these industries (compare Appendix C, Table C.1).[17] Although this classification of the capital goods sector and trade is quite commonplace, it should be noted that the production and trade data also comprise some durable consumer goods.[18]

In the reference estimations, we represent the user cost of domestically produced capital goods (P_{Kd}) by a dollar index of production prices of the domestic capital goods industries (1980 = 1); thus the stock is evaluated at replacement cost.[19] The border price of imported capital

14 Marquez (1984, 1985) assumes, for example, that the developing countries have a capital-output ratio of 3.38. If a constraint is placed on the production function, namely that capital goods are the only factor of production and that some fixed capital-output ratios prevail in the economy, the capital stock *and* the depreciation rate can be recovered from investment data (Dadkhah and Zahedi 1986).

15 Compare, e.g., Farzin (1988), Field (1987), and Park (1985).

16 The beginning year capital stock of the year 1970 (K_{70}) is derived as

$$K_{70} = .125\, I_{63} + .25\, I_{64} + .375\, I_{65} + .5\, I_{66} + .625\, I_{67} + .75\, I_{68} + .875\, I_{69} + 1.0\, I_{70},$$

where I_t represents real investment in year t.

17 Only the trade with the developed market economies is taken into account here (compare Appendix C) in order to enable data coverage until 1987 for all countries. These data should cover a very high share of total capital goods trade of the developing countries.

18 Estimation results would be much more precise, if sector-specific production functions could be analyzed. Although data on output, employment, and sometimes even total investment are collected on an industry level (by UNIDO), data on the use of imported products in the different industries is not available in a form suitable for time-series estimation. Only very few countries incorporate the destination of imported means of production in input-output tables. Unfortunately, these input-output matrices are published with a considerable time lag and are most often not collected on an annual basis.

19 The replacement cost concept is also employed by Bautista (1980), König (1990), and Michalopoulos (1975).

goods (dollar index 1980 = 1) is adjusted for by the implicit charge rate on capital goods as described above (Section 5.2.2.) to provide a proxy for the user cost of imported capital goods (P_{Kf}).

5.3.2.2. Estimation of the regression parameters

The log-linear specification of equation 5.3 is first estimated applying ordinary least squares. Since a lagged endogenous variable is included on the right hand side of the equation, the usual test for first-order autocorrelation (Durbin-Watson test) is not applicable. Instead the Durbin-*h* statistic (denoted as D*h* in the tables below) is computed,[20] defined as

$$(5.4) \qquad Dh = r \left[\frac{T}{1 - T*Var(ß)} \right]^{.5},$$

where r is the estimate of the true first-order autocorrelation from the regression of u_t on u_{t-1}, T the number of observations, and Var(ß) the variance of the estimated parameter of the lagged endogenous variable. The *h*-statistic and the null-hypothesis of no first-order autocorrelation are approximately normally distributed with unit variance, so that the critical value of rejecting the null-hypothesis at the 5% level becomes 1.645. In those cases where T*Var(ß) happens to be larger than 1, the *h*-statistic cannot be calculated. Instead, the first-order Lagrange Multiplier (LM) test is conducted, which Durbin (1970) has shown to be an asymptotically equivalent procedure.[21]

If first-order autocorrelation is detected,[22] the parameter estimates are biased and inconsistent.[23] A two-stage estimating procedure has to be applied. We conduct an instrumental variable (IV) regression of the dependent variable on lagged values of the exogenous variables and use the lagged predicted values of this regression as an instrument for the lagged endogenous

---

20 This test is due to Durbin (1970).

21 A regression is run of the residuals obtained from OLS on lagged residuals, the lagged dependent variable, and all other independent variables. If the coefficient of the lagged error term is significantly different from zero, the null-hypothesis of no first order autocorrelation is rejected. See Johnston (1984, p.318) and Spanos (1986, pp.518-519) for a description of the LM-test.

22 Since the number of observations is relatively small, Lagrange-Multiplier tests on higher-order autocorrelation are not conducted.

23 Compare, e.g., Pindyck and Rubinfeld (1981) p.193.

variable.[24] It is then possible to correct for the first-order autocorrelation and apply generalized least squares (GLS). In order to save the first observation of the relatively small sample, the Prais-Winsten method is employed (Prais and Winsten 1954). In the cases in which an adjustment for first order autocorrelation has to be performed both the statistic referring to the initial regression *without* adjustment (LM or D*h* statistics) as well as the estimate of the coefficient of autocorrelation (rho) are presented.

The parameter of most interest is the short run elasticity of substitution a_1. The assumption of constancy over time of this variable is operational and, for example, policy changes might have an impact on the elasticity of substitution. In order to test the stability of the parameter over the period considered, a *separate* regression is run including two slope-shifting dummy variables, which attempt to represent the influence of policy changes on the parameter estimate of the short run elasticity of substitution.

One dummy variable is constructed to encompass the trade policy change immediately following the outbreak of the debt crisis (in 1982). Laird and Nogués (1988a and 1988b) note that all highly indebted countries reacted to the payments crisis by increasing import barriers. But the tightening of the trade policy was subsequently reversed from 1985 onwards, since a widespread liberalization of both tariff and quantitative restrictions occurred (Laird and Nogués 1988a and 1988b, UNCTAD 1989, Chapter 5, Whalley 1989). The second dummy relates to this period (1985-1987). The first dummy variable assumes the value zero until 1982 and is then equal to the relative price variable; the second dummy variable consists of the value zero until 1984, while being equal to the relative price term from 1985 to 1987. A separate regression is run which includes both of these additional exogenous variables. An F-test is performed to examine whether the two dummies add significantly to explaining the variance of the endogenous variable. These regression results are not presented as they mainly serve to gauge whether estimated elasticities are stable or should rather be viewed as average values over the estimation period.

With regard to data on imported capital goods obtainable from 1963 onwards, proxies for the capital stocks' ratio are available from 1970 and the inclusion of the lagged endogenous variable restricts the possible starting year of estimation to 1971. But most countries report import duties only from 1972 and this figure is necessary to calculate the protection proxy. Beginning year values of the estimation have to be restricted to 1972.[25] We present estimates of the regres-

---

24 This method is described in Koutsoyiannis (1977, p.323). Compare also Arize (1990, pp.895-896). The instrumental variable estimator yields only asymptotically unbiased estimates.

25 Production data for Mexico in the year 1987 were incomplete, so that the estimation period is confined to the 1972-1986 period.

Table 5.5 : Reference estimations for capital goods, linear specification (1972-1987)[1]

Country	Estimator	a_0	a_1	a_2	a_3	R^2_{adj}	Serial	stability (F-test)
Argentina	OLS	.3579 (.2967)	.3757*** (.0705)	-.0127* (.0066)	.7952*** (.0924)	.9437	D*h* = .7011	2.643
Brazil	IV, GLS	.3257 (.5569)	-.0277 (.0292)	.0094 (.0264)	.8069* (.4234)	.9870	D*h* = 2.567 rho = -.5332	.1878
Chile	OLS	.1799*** (.0503)	.3037*** (.0536)	-.0240*** (.0037)	.7641*** (.0539)	.9809	D*h* = .5921	.9120
Columbia	OLS	.0695** (.0281)	.3037*** (.0406)	-.0030 (.0019)	.7606*** (.0586)	.9362	D*h* = 1.292	2.839
Ecuador	OLS	-.2883 (.3287)	.0266 (.0726)	.0169 (.0142)	.7841*** (.2095)	.9749	D*h* = .7951	2.011
Indonesia	OLS	-1.082*** (.2771)	.2986*** (.0527)	.0555*** (.0136)	.4430*** (.1249)	.9926	D*h* = .3971	.0014
Malaysia	OLS	-.0205 (.0305)	.4240*** (.1395)	-.0176*** (.0047)	.7492*** (.0827)	.9723	D*h* = -.1017	1.096
Mexico[2]	IV, GLS	.2241 (.3481)	.1762* (.0988)	-.0119 (.0122)	.7329** (.3340)	.9272	D*h* = 1.665 rho = -.3359	11.52***
Peru	OLS	.1344*** (.0347)	.2134* (.1036)	-.0051 (.0046)	.7297*** (.0948)	.9070	D*h* = .4453	2.547
Philippines	OLS	-.0490 (.0379)	.0924 (.0839)	-.0068** (.0022)	.7300*** (.1187)	.8311	D*h* = .5545	.7113
Uruguay	OLS	.2361* (.1201)	.3365*** (.0740)	-.0105** (.0040)	.7772*** (.0838)	.9406	D*h* = .6106	1.960
Venezuela	OLS	-.2182** (.0847)	.1323** (.0516)	.0234** (.0078)	.2879 (.2267)	.9396	D*h* = .4937	.0806

1 See text for a description of the variables included. Standard deviations of the parameter estimates are in brackets. If the instrumental variable estimator is applied, estimates of standard deviations of the parameters are only asymptotically unbiased.

2 Estimation period 1972-1986.

*** Parameter estimate is different from zero (and the null-hypothesis of the stability test is rejected) at the 1% level of significance.

** Parameter estimate is different from zero (and the null-hypothesis of the stability test is rejected) at the 5% level of significance.

* Parameter estimate is different from zero (and the null-hypothesis of the stability test is rejected) at the 10% level of significance.

sion parameters in Table 5.5. Standard deviations of the parameters are shown in brackets below the estimates with stars indicating the significance level. The coefficient of determination adjusted for the degrees of freedom is presented ('R^2_{adj}'), as well as the test for first-order autocorrelation ('serial').

Except for Brazil, the parameter estimates of the short run elasticity of substitution (a_1) carry the expected positive sign. The Brazilian result will be discussed below. Only in two of the eleven remaining cases is the short run elasticity not significant, namely for Ecuador and the Philippines. Estimates for the two countries are relatively small (.027 in the Ecuadorian case and .092 for the Philippines). The estimate for Ecuador especially has to be crosschecked in the modifications below since it is possible that multicollinearity has affected the variances.[26] The other countries show parameter estimates significant at the 1% level (Argentina, Chile, Columbia, Indonesia, Malaysia, Uruguay), at the 5%-level (Venezuela), and at the 10%-level (Mexico and Peru). The estimated short run elasticity is highest for Malaysia (.424), being followed by Argentina, (.376) and Uruguay (.337). All estimates are significantly different from 1, indicating that the two capital stocks are complements in the short run.

The computed coefficients of the lagged endogenous variable (a_3) are in the expected range between 0 and 1. They are significant for all countries with the exception of Venezuela. The coefficient is highest (with a correspondingly low speed of adjustment λ_1 of the capital stock ratio) for Argentina, Brazil, Ecuador, and Uruguay. On the other hand, Indonesia and Venezuela appear to possess the shortest adjustment time; in both cases more than half of the desired adjustment in the capital stock ratio takes place in the year of the relative price change.

Turning to the test of the stability of the parameter a_1 (short run elasticity of substitution), which we perform in a separate regression including the two slope shifting dummy variables, the null-hypothesis that no structural break occurred is accepted for all countries except for Mexico. Overall, it emerges that the short run elasticity remained stable during the period under consideration, irrespective of major policy changes in the 1980s.

5.3.2.3. Estimation of the structural parameters

In order to obtain estimates not only of the regression parameters but also of the structural coefficients, we incorporate the non-linear parameter combinations in the specification of equation 5.3 and apply non-linear least squares. This is solely a reordering of parameters; the additive

---

26 While the adjusted coefficient of determination is high, only the coefficient of the lagged endogenous variable (a_3) is significantly different from zero.

Table 5.6:
Reference estimation for capital goods, non-linear specification (1972-1987)[1]

Country	λ_1	σ_k	$(z_{Kf}-z_{Kd})$	a
Argentina	$.2048^{**}$	1.835^{*}	.0746	$.7216^{***}$
	(.0924)	(.9001)	(.0794)	(.1691)
Chile	$.2359^{***}$	1.287^{***}	.3533	$.6439^{***}$
	(.0539)	(.3650)	(.4129)	(.0484)
Columbia	$.2395^{***}$	1.268^{***}	.0468	$.5569^{***}$
	(.0586)	(.3620)	(.0647)	(.0271)
Ecuador	.2156	.1233	$.0894^{*}$	.0001
	(.2095)	(.3105)	(.0408)	(.0005)
Indonesia	$.5570^{***}$	$.5361^{***}$	$.2146^{***}$	.0259
	(.1249)	(.1565)	(.0689)	(.0291)
Malaysia	$.2508^{***}$	1.690^{**}	.1015	$.4879^{***}$
	(.0827)	(.7357)	(.1056)	(.0178)
Mexico[2]	.2671	.6597	-.1307	$.7810^{*}$
	(.3340)	(1.014)	(.4259)	(.3975)
Peru	$.2703^{**}$	$.7896^{**}$	-.0880	$.6525^{***}$
	(.0948)	(.3491)	(.2085)	(.0774)
Philippines	$.2700^{**}$	.3421	-.0385	$.3704^{**}$
	(.1187)	(.3908)	(.0323)	(.1334)
Uruguay	$.2228^{**}$	1.510^{**}	.0927	$.6685^{***}$
	(.0838)	(.6446)	(.1106)	(.0899)
Venezuela	$.7121^{***}$	$.1858^{**}$	$.0404^{***}$	.1612
	(.2267)	(.0716)	(.0053)	(.0918)

1 See text for a description of the variables included. Asymptotic standard deviations of the parameters are in brackets. The adjusted coefficient of determination (R-square) is the same as in the linear estimation. The same holds for the Durbin-h statistic. Non-linear parameter estimation is not applied to the Brazilian data because the estimate of the short-run elasticity of substitution turned out negative in the linear estimation.

2 Estimation period 1972-1986.

*** Parameter estimate is different from zero at the 1% level of significance.

** Parameter estimate is different from zero at the 5% level of significance.

* Parameter estimate is different from zero at the 10% level of significance.

specification of the error term in equation 5.3 is not altered so that both statistics of the coefficient of determination and the autocorrelation test statistic (the parameter a_3 is a linear function of the structural parameter λ_1) remain the same as presented in Table 5.5.[27] The Brazilian data are not subjected to non-linear estimation since the short run elasticity of substitution carries the wrong sign in the linear estimation. Estimates for the other countries are presented in Table 5.6.

As expected, those economies that exhibit a short run elasticity estimate not significantly different from zero also carry low and insignificant values of the long run elasticities (Ecuador and the Philippines). Argentina, Chile, Columbia, Malaysia, and Uruguay show long run values above 1, although in none of them is the estimate significantly different from 1. The estimate of σ_k for Peru is significant and below unity, but - again - not significantly different from it. Although the short run elasticity estimate for Mexico is different from zero in Table 5.5 at the 90% confidence level, this does not hold for the long run elasticity estimate.[28] All signficant estimates of the difference in the rates of technological progress (z_{Kf}-z_{Kd}) are positive (Ecuador, Indonesia, Venezuela), implying that imported machinery and equipment are subject to higher rates of technological progress than indigenously manufactured capital goods. While the estimates for Ecuador and Venezuela are relatively low, it is very high in the Indonesian case. The same holds for Chile, although the estimate is insignificant in this instance.[29]

5.3.3. Modification 1: the depreciation rate

Estimates of the capital stocks which are generated by accumulating real net investment per year vary directly with the assumed life-span of additions to the stock. A relatively wide dispersion of depreciation rates for developing countries are put to use in the literature.While Bergsten et al. (1985) implicitly assume an average rate of 20%,[30] Kell and Marchese (1991)

---

27 The resulting estimates of the structural coefficients are only asymptotically efficient and consistent. The t-statistics conducted on these non-linear parameter estimates relate to large samples only. See Gallant (1987), pp.16-19.

28 This is due to the circumstance that the adjustment parameter λ_1 is not significantly different from zero.

29 It is possible that the time trend picks up other influences than disembodied technological progress as assumed here. For example, if the depreciation rates of the two different capital stocks are not equal, this effect might be picked up by the time trend. Lucas (1989, p.158) says that if the estimate of the rate of technological progress is not significant, "this is not to deny the existence of technical progress altogether, as fixed capital investment may well embody improved technologies".

30 Bergsten et al. (1985) represent their capital stock variable of developing countries by five years' cumulation of gross fixed investment (p.208). More precisely, they assume no diminution of value for five years and then a sudden 100% depreciation.

employ 12.5% as in the reference estimations of this study. Field (1987, p.1285) and Ikemoto (1986, p.387) apply a rate of 10%, while Erbe (1982, p.127) and the IMF (1989, p.66) posit a rate of 5%.

In order to trace the influence of the depreciation rate on the estimates, we recalculate capital stock values applying a linear depreciation rate of 10% with no diminution of value in the period of installation. Observations of the capital stocks are then available only from 1972 and the inclusion of the lagged endogenous variable confines the starting year of estimation to 1973, reducing the already small number of degrees of freedom to eleven. Results of reestimating equation 5.3 are presented in Table 5.7.

The overall influence of the lower depreciation rate appears to reduce the short and long run elasticity estimates in absolute terms, although only marginally for most countries. None of the estimates of a_1 is different from the numerical value obtained in the reference case at the 5% significance level (using the error term of modification 1 results). The estimate of a_1 is higher than in the reference case for Ecuador and Malaysia and lower for all other countries. Concerning the significance levels of the estimates, the calculations of the reference case are broadly reproduced. The stability (F-test) statistic in Table 5.7 indicates that the short run elasticity of substitution changed only for Mexico in the 1980s.

The long run estimates of the elasticity of substitution obtained from applying non-linear least squares to equation 5.3 (based on the changed depreciation rate) are also included in the table (σ_k). Results are again similar to the reference scenario. If a test is performed on whether the new estimates are equal to those calculated in the reference case, only one country exhibits a parameter value different from the previous one at the 5% level of significance, namely Malaysia.[31] But the estimates for Malaysia are not significantly different from 1 in either the reference regression or the one with the lower depreciation rate.

5.3.4. Modification 2: an alternative formulation of the user cost of capital

Theoretically, several factors enter into the user cost of capital goods: (a) the price of capital goods, (b) the interest rate, (c) the real depreciation rate, (d) the tax rate on profits, (e) the depreciation allowance (tax), and (f) the expected capital gains.[32] It is difficult to incorporate all of the above influences in a measurable user-cost concept. Data on specific characteristics of the

---

31 The error terms of the modification 1 estimations are used for the test.

32 See for a discussion of user costs concepts Coen (1971, pp.157-163) and Mirakhor and Montiel (1987, p.69).

Table 5.7 : Modification 1 (depreciation rate 10%) estimations for capital goods (1973-1987)[1]

Country	Estimator	a_0	a_1	a_2	a_3	σ_k	R^2_{adj}	Serial	stability (F-test)
Argentina	OLS	.4584* (.2514)	.2936*** (.0500)	-.0147** (.0048)	.7863*** (.0790)	1.374** (.5804)	.9626	D*h* = .2275	1.976
Brazil	IV, GLS	.1896 (.5420)	-.0243 (.0212)	.0052 (.0241)	.8983* (.4120)	-[2]	.9957	D*h* = 5.635 rho = -.4143	.9311
Chile	OLS	.1860** (.0637)	.2263*** (.0425)	-.0254*** (.0047)	.7427*** (.0678)	.8797** (.3102)	.9867	D*h* = -.9946	1.087
Columbia	OLS	.0959*** (.0246)	.2373*** (.0337)	-.0046** (.0016)	.7460*** (.0550)	.9342*** (.2506)	.9453	D*h* = .8122	1.883
Ecuador	OLS	-.4567 (.3722)	.0466 (.0588)	.0224 (.0153)	.6778** (.2352)	.1445 (.1754)	.9814	D*h* = 1.452	1.586
Indonesia	OLS	-1.193*** (.3726)	.2816*** (.0582)	.0610*** (.0178)	.3672* (.1805)	.4450*** (.1294)	.9925	D*h* = .5955	.0256
Malaysia	OLS	-.0519** (.0203)	.4587*** (.0866)	-.0298*** (.0045)	.5687*** (.0733)	1.064*** (.2277)	.9892	D*h* = -.9246	1.568
Mexico[3]	IV, GLS	.2317 (.3711)	.1433 (.0894)	-.0123 (.0136)	.7147* (.3725)	.5024 (.8160)	.9255	D*h* = 1.711 rho = -.3285	8.259**
Peru	OLS	.1224*** (.0284)	.1915** (.0836)	-.0037 (.0044)	.7218*** (.0930)	.6883** (.2727)	.9369	D*h* = .1079	1.919
Philippines	OLS	-.0483 (.0446)	.0785 (.0822)	-.0052* (.0024)	.7785*** (.1637)	.3545 (.5433)	.7876	D*h* = .9666	.8438
Uruguay	OLS	.3798** (.1593)	.2854*** (.0657)	-.0178*** (.0054)	.6672*** (.1069)	.8576** (.3445)	.9578	D*h* = .2169	2.905
Venezuela	OLS	-.1989** (.0765)	.1111** (.0408)	.0222** (.0073)	.2790 (.2362)	.1540** (.0604)	.9528	D*h* = .1246	.0216

1 See text for a definition of the variables included. Standard deviations of the parameter estimates are presented in brackets. The estimate of the standard deviation of the long-run elasticity of substitution is only asymptotically unbiased. The same holds for the estimates of the standard deviations if the IV estimator is used.

2 The estimate of the long-run elasticity of substitution is not computed as the short-run value has the wrong sign.

3 Estimation period 1973-1986.

***, **, * Parameter estimate is different from zero (and null-hypothesis of the stability test is rejected) at the 1%, 5%, 10% level of significance.

Table 5.8 : Modification 2 (user cost concept) estimations for capital goods (1973-1987)[1]

Country	Estimator	a_0	a_1	a_2	a_3	σ_k	R^2_{adj}	Serial	stability (F-test)
Columbia	OLS	.0442 (.0312)	.3113*** (.0540)	.0031 (.0025)	.8061*** (.0646)	1.605** (.6255)	.9232	D*h* = 1.129	.3357
Ecuador	OLS	-.2856 (.3109)	-.0473 (.0937)	.0155 (.0129)	.7801*** (.2035)	-[2]	.9760	D*h* = .3587	1.189
Mexico[3]	OLS	.1362 (.1033)	.2024*** (.0308)	-.0020 (.0038)	.7641*** (.1026)	.8575*** (.1264)	.9843	D*h* = .6542	.8749
Philippines	OLS	-.0694 (.0436)	.0816 (.0672)	-.0074*** (.0020)	.5778*** (.1399)	.1933 (.1295)	.7041	D*h* = .8639	.3513
Venezuela	IV, GLS	-.2458 (.1405)	.1204* (.0654)	.0280 (.0136)	.1454 (.3879)	.1409* (.0697)	.9221	LM = 3.329* rho = -.3309	.8333

1 See text for a definition of the variables included. Standard deviations of the parameter estimates are presented in brackets. The estimate of the standard deviation of the long-run elasticity of substitution is only asymptotically unbiased. The same holds for the estimates of the standard deviations if the IV estimator is used.

2 The estimate of the long run elasticity of substitution is not computed as the short run value has the wrong sign.

3 Estimation period 1973-1986.

*** Parameter estimate is different from zero (and the null-hypothesis of the LM is rejected) at the 1% level of significance.

** Parameter estimate is different from zero (and the null-hypothesis of the LM is rejected) at the 5% level of significance.

* Parameter estimate is different from zero (and the null-hypothesis of the LM is rejected) at the 10% level of significance

determinants of the user costs are difficult to generate (like real depreciation rates on a macroeconomic level and the effective tax rates).

In the reference case and modification 1, the nominal price of investment goods was used as a proxy for total user costs (including the protection proxy for imported capital goods). The existing stock was evaluated at replacement costs. Now we apply an alternative user cost concept. Following Marquez (1985, p.44) the rental price of the domestic capital stock (measured as a dollar index) at time t ($P_{Kd,t}$) is defined as

$$(5.5) \qquad P_{Kd,t} = E\left\{ P^{n}_{Id,t-1} \left[r_t + \delta_d \frac{P^{n}_{Id,t}}{P^{n}_{Id,t-1}} \right] \right\},$$

where r_t represents the nominal interest rate, P^{n}_{Id} the price index of domestically manufactured investment goods (in domestic currency), δ_d the depreciation rate on the domestic capital stock assumed to be 12.5%, and E the nominal exchange rate index.[33] Similarly, for foreign capital goods a user cost index P_{Kf} is constructed according to

$$(5.6) \; P_{Kf,t} = P'_{MCap,t-1}(1+t_{Cap,t-1}) \left[r^{*}_{t} + \delta_f \frac{P'_{MCap,t}\,(1+t_{Cap})}{P'_{MCap,t-1}\,(1+t_{Cap,t-1})} \right],$$

where r_t^* is the (average) nominal interest rate on foreign loans, P'_{MCap} the price index of imported capital goods, δ_f the depreciation rate on the capital stock made up of imports (also assumed equal to 12.5%), and t_{Cap} the tariff rate on imported machinery and equipment. The rental price concept in equations 5.5 and 5.6 is similar to the one developed by Coen (1971).

Data on nominal lending rates in developing countries covering the period under consideration are rare. For five countries (Columbia, Ecuador, Mexico, the Philippines and Venezuela) interest rate data could be obtained from various sources (see Appendix C.4.3.). Table 5.8 reports the regression results.

Compared to the use of simple price indices, the inclusion of the interest and depreciation rate in the user cost concept of the two investment goods does not alter the results significantly.[34]

---

33 Nominal interest rates in developing countries are most often regulated. The availability of domestic credit might be a better indicator of the cost of domestic funds than the nominal interest rate.

34 The estimation period is confined to the 1973-1987 period. One observation is lost due to the inclusion of the lagged tariff rate in equation 5.6.

The Ecuadorian estimate of a_1 turns negative but it remains insignificant. The most important difference emerges for Mexico: The incorporation of the alternative user cost concept changes the short run estimate of the elasticity of substitution only marginally, but it is now highly significant at the 1% level while it only was significant at the 10% level in the reference estimation. No autocorrelation is detected in the estimation with the different user cost concept (as was the case beforehand), the long run elasticity estimate is highly significant, and the stability test on the parameter a_1 signifies no structural break in the 1980s.

5.3.5. Modification 3: using gross domestic investment data

5.3.5.1. Conceptual aspects of the data

The third sensitivity analysis is concerned with the measurement of the domestic capital stock. The production of industries classified as producing capital goods was used as a measure of stock formation from domestically produced capital goods in the reference calculations. This proxy has several shortcomings:

(a) It assumes that all goods produced in these industries are indeed capital goods. But a share of output consists also of some durable consumer goods.

(b) The industry classification of each firm (whether it belongs either, e.g., to the metal products industry or to the iron and steel industry) takes place according to the nature of its dominant production. The capital goods produced by an industry manufacturing mainly consumer goods and own-account production of investment goods are hence not incorporated in the capital goods measure employed so far.

(c) The proxy does not take into account changes in the inventories of investment goods.

In order to crosscheck the estimates derived from the production statistics, we use a different *type* and *source* of data to construct a series for the domestic capital stock. These are gross domestic investment (GDI) figures originating on the expenditure side of national income accounting supplied by the World Bank (1989d). Gross domestic *fixed* investment is "made up of all outlays (purchases and own-account production) of industries, producers of government services, and producers of private nonprofit services on additions of new and imported durable goods to their stocks of fixed assets, reduced by the proceeds of net sales (sales less purchases) of similar secondhand and scrapped goods" (World Bank 1988b, p.6).[35] GDI is then defined as the sum of gross fixed investment and the change in inventories. All three measurement problems associated with the proxy originating from production statistics do not apply here. Inventory changes

---

35 Outlays on capital goods for military purposes are excluded.

are taken into account, own-account production is included, and the production of consumer-durables is by definiton excluded. But the expenditure data do include a higher share of government investment - such as infrastructural expenditures on the transportation, health, and educational system - than the production figures.[36] The capital stock denoted as K_d in equation 5.3 now encompasses own-account production outlays of governmental service enterprises for, e.g., road-making or construction of hospitals. One can anticipate that the bundle of domestically produced investment goods shows a lower substitution elasticity versus the imported capital goods than when production figures of capital goods industries were taken into account.[37]

5.3.5.2. Data generation

If expenditure data on gross domestic investment (GDI) are used, they include outlays on domestically produced capital goods as well as on imported ones. The nominal dollar value of GDI ($P_{GDI}GDI$) becomes

$$(5.7) \qquad P_{GDI}GDI = P_{Id}I_d + P'_{MCap}\,(1+tCap)\,M_{Cap}$$

where P_{Id} represents the dollar price index of investment originating domestically (I_d), whereas P'_{MCap} and M_{Cap} stand for the price and purchase of imported capital goods. A change of the dollar price index of GDI comprises the effect of a change of P_{Id}, t_{Cap}, and of P'_{MCap}. Domestically originating investment I_d and its price P_{Id} cannot be observed directly. We assume that P_{Id} moves in line with the dollar price index of domestically produced capital goods derived from the production statistics. Real dollar expenditures on domestically produced investment goods (I_d) can then be computed from equation 5.7,[38] which serves to calculate the domestic capital

---

36 The 'production measure' used in the reference case includes government orders of capital goods and also the output of government owned enterprises in the capital goods sector, but the share of government activity in total investment is substantially higher in the expenditure data on GDI, since infrastructural investment is also included.

37 Additionally, Marquez (1985, p.52, fn 11) mentions that the government may use a social criterion and not market prices in evaluating the desirability of particular investment projects. It has also been observed that in times of budget stringency, investment expenditures are cut overproportionally relative to current expenditures (Mirakhor and Montiel 1987, pp.69-70). Further, as the government is often a large demander and producer of capital goods (especially in regard to infrastructural investment) and at the same time sets tariffs and quantitative restrictions on imported machinery and equipment, the assumption that it faces given prices of the capital goods might not be valid any more. These arguments lead to the use of the production data in the reference estimation.

38 This calculation of I_d implies that inventory changes take place only for domestically produced investment goods. The calculation is only possible if an assumption is made about the value of t_{Cap} before 1972. The value of t_{Cap} in the 1963 to 1971 period is set equal to the 1972 value. This has only an effect on the estimate of the initial capital stock of K_d.

Table 5.9 : Modification 3 (expenditure data on GDI) estimations for capital goods (1972-1987)[1]

Country	Estimator	a_0	a_1	a_2	a_3	σ_K	R^2_{adj}	Serial	stability (F-test)
Argentina	OLS	.6490** (.2812)	.2925*** (.0563)	-.0226*** (.0061)	.7742*** (.0791)	1.296** (.5391)	.9721	D*h* = .1709	1.478
Brazil	OLS	.6858** (.2696)	.3875*** (.0472)	.0258*** (.0056)	.6580*** (.1216)	1.133*** (.3302)	.9855	D*h* = 1.439	8.585***
Chile	IV, GLS	.5464 (.7889)	.0615 (.0642)	-.0102*** (.0031)	.6689 (.5300)	.1857 (.4337)	.5287	D*h* = 4.745 rho = .4016	.1782
Columbia	OLS	.7553*** (.2063)	.3436*** (.0700)	.0035 (.0025)	.5063*** (.1300)	.6960*** (.1565)	.8514	D*h* = .9928	1.017
Ecuador[2]	OLS	-.0286 (.1205)	-.0718 (.0958)	.0108** (.0044)	.9678*** (.1255)	-[3]	.9070	D*h* = .-5446	1.955
Indonesia	OLS	.5778*** (.1796)	.3319*** (.0831)	.0189*** (.0049)	.5720*** (.1269)	.7755*** (.2029)	.9405	D*h* = 1.420	9.389***
Malaysia	IV, GLS	.7088 (.6915)	.4075 (.2942)	-.0158 (.0219)	.6139* (.3391)	1.056 (.9996)	.7224	D*h* = 5.006 rho = -.6485	.0719
Mexico[4]	OLS	.5486 (.3064)	.1025* (.0570)	-.0027 (.0021)	.7378*** (.1532)	.3910 (.3349)	.6134	D*h* = .4809	12.09***
Peru	OLS	.2264 (.1628)	.0745 (.0678)	-.0136*** (.0040)	.8089*** (.1199)	.3903 (.4975)	.9516	D*h* = 1.181	3.043*
Philippines	OLS	.2354 (.1406)	.0609 (.0710)	-.0008 (.0037)	.8804*** (.1015)	.5093 (.8576)	.9651	D*h* = .1751	.7681
Uruguay	OLS	.6656** (.2480)	.2274** (.0965)	-.0131*** (.0041)	.7230*** (.1041)	.8209* (.4001)	.9118	D*h* = 1.571	.1971
Venezuela	IV, GLS	1.308** (.5182)	.1369** (.0541)	-.0165** (.0072)	.3006 (.2742)	.1957* (.1002)	.9088	D*h* = 2.429 rho = -.3893	.4855

1 See text for a definition of the variables included. Standard deviations of the parameter estimates are presented in brackets. The estimate of the standard deviation of the long-run elasticity of substitution is only asymptotically unbiased. The same holds for the estimates of the standard deviations if the IV estimator is used.

2 Estimation period 1973-1987.

3 The estimate of the long-run elasticity of substitution is not computed as the short-run value has the wrong sign.

4 Estimation period 1972-1986.

***, **, * Parameter estimate is different from zero (and the null-hypothesis of the stability test is rejected) at the 1%, 5% and 10% level of significance.

stock K_d applying a depreciation rate of 12.5%. The definition of imports in terms of the Standard International Trade Classification is different from the reference estimation as well, since now consumer durables are excluded.[39]

5.3.5.3. Regression results

Estimates of the parameters are shown in Table 5.9. Considering first the results for the short run elasticity of substitution (a_1), they are similar to the reference case of Section 5.3.2. for most countries, namely for Argentina, Columbia, Ecuador[40], Indonesia, Mexico, the Philippines, Uruguay, and Venezuela. In none of these cases is the estimate different at the 5% significance level from the value obtained in the reference case (using either the error term of the parameter estimate of the modification estimation or that of the reference case parameter estimate). It should be noted that the estimate of a_1 for Malaysia is close to the one obtained in the original estimation (.4075 vs. .4240), but that the parameter value is now not significantly different from zero. It is possible that multicollinearity affected the results. A relatively high coefficient of determination (R^2_{adj} =.7224) is present, but only one estimated coefficient is different from zero at the 10% significance level in the Malaysian regression. Additionally, while the error statistic did not indicate the presence of autocorrelation in the reference estimation, first order autocorrelation is now detected. The Chilean and Peruvian short run elasticity estimates have the correct sign, but are now insignificantly different from zero. This result can be motivated by the theoretical consideration that the expenditure measure on investment includes a much higher governmental share than the production figure made use of in the reference scenario. Indeed, eight of the twelve estimates of a_1 and seven estimates of σ_k are lower compared to the analysis of production data (compare Table 5.9 with Tables 5.5 and 5.6).

The most noticeable difference can be traced for Brazil. The short run estimate turned out with the wrong sign when the production figures were used for the domestically produced capital goods in Table 5.5; in this modification the correct sign appears, which is highly significant and among the largest short run elasticities observed in the sample. The same holds for the long run estimate (σ_k). Brazil commands a developed capital goods industry, producing a wide variety of investment goods. One would have expected a result similar to the one observed in Table 5.9. Further, as pointed out above, it is reasonable to anticipate that the estimate of a_1 is lower, when

39 Appendix C, Table C.2 defines imported capital goods in terms of the SITC definitions as used here. The price of the redefined imported capital goods is again obtained by aggregating nominal and real import categories separately and deriving the implicit price deflator.

40 Data on investment expenditure for Ecuador were only available from 1965 on so that the starting period of estimation is confined to 1973.

expenditure data are used to calculate K_d than if production figures are used. One potential explanation of this result might be that a measurement error is present in the industrial statistics, which would cause estimates of both the regression and structural parameters to be biased and inconsistent. In the final modification of equation 5.3 in the next sub-section, expenditure data on domestic investment goods are used instead of production data for Brazil.

The F-test shown in Table 5.9 is arranged in the same way as in previous tables and measures the stability of the short run elasticity of substitution between the two capital stocks in the 1980s. The null-hypothesis that a structural break was absent cannot be rejected for Brazil, Indonesia, Mexico, and (at the 10% level of significance) for Peru.

5.3.6. Modification 4: modelling the foreign exchange impact

This modification employs the government reaction function described in equation 5.2 instead of using the protection proxy. Two issues are of interest: (a) whether the estimates of the short and long run elasticities of substitution differ significantly from the reference case in which the protection proxy was included, and (b) whether the availability of foreign exchange can be shown to have a significant influence on the internal price of imported capital goods. The results are sensitive to the assumption that the variability of protectionist policies is related to the availability of foreign exchange. If a functional relationship between foreign exchange availability and pricing policies of imported capital goods is not present, but the protectionist policies varied over the period considered, results between the two different approaches must not be equal.

5.3.6.1. The estimating equation

Analogous to equation 5.1, we express the user cost of the imported capital stock (P_{Kf}) as the border price of imported capital goods (P'_{MCap}) times the effect of tariff and non-tariff barriers on imported capital goods:

$$(5.8) \qquad P_{Kf} = P'_{MCap}\left[1+t_{MCap}\right].$$

We now assume that $(1+t_{MCap})$ can be represented as a function of foreign exchange availability (FX_{Cap}):

$$(5.9) \qquad (1+t_{MCap}) = f_0 \, FX_{Cap}^{f_1}$$

with f_1 as the foreign exchange elasticity of protectionist policies on capital goods and $f_1 \leq 0$. If foreign exchange is abundant, the term on the right hand side tends to f_0, the lower bound of the protectionist effect which is invariant to foreign exchange availability.

Combining equations 4.18, 5.8, and 5.9 yields[41]

$$(5.10) \frac{K_d}{K_f} = \left[\frac{a}{1-a}\right]^{\sigma_k\lambda_1} \left[\frac{P'_{MCap}}{P_{Kd}}\right]^{\sigma_k\lambda_1} e^{(z_{Kf}-z_{Kd})t\lambda_1\sigma_K\theta} \left[\frac{K_{d,t-1}}{K_{f,t-1}}\right]^{1-\lambda_1} f_0^{\sigma_k\lambda_1} FX_{Cap}^{f_1\sigma_k\lambda_1}.$$

Assuming a multiplicative error of the exponential form in equation 5.10 and taking logarithms results in

$$(5.11) \ln\left[\frac{K_d}{K_f}\right] = a'_0 + a'_1 \ln\left[\frac{P'_{MCap}}{P_{Kd}}\right] + a'_2 t + a'_3 \ln\left[\frac{K_{d,t-1}}{K_{f,t-1}}\right] + a_4 \ln(FX_{Cap}) + u_{11}$$

with

$$a'_0 = \sigma_k\lambda_1 \left\{ \ln\left[\frac{a}{1-a}\right] + \ln(f_0) \right\}$$

---

41 Alternatively, one can model the adjustment time as a function of the foreign exchange availability as proposed by Khan and Knight (1986) and Marquez (1984, 1985) (compare Section 3.3.2.). The adjustment parameter, λ_1, could then be included in the estimating formulation as

$$\lambda_1 = \lambda_0 + f_2 \, e^{(-FX)},$$

with $-\lambda_0 \leq f_2 \leq 0$ and $0 < \lambda_0 \leq 1$. λ_0 represents the upper bound of the speed of adjustment which is reached only if foreign exchange availability is very large. Inclusion of this function in addition to the trade policy function (equation 5.9) produces multicollinearity problems, as the same variable is included twice (although once in logs and once in an exponentional form and thus not linearly dependent). As argued in Chapter 3, preference is given to the price formulation, since a prolonged adjustment time does have an influence on the opportunity cost of an investment as well. Further, most trade barriers (surcharges, foreign exchange taxes, auctioning of licenses, tariffs etc.) have a direct price impact.

$$a_1' = \sigma_k \lambda_1$$

$$a_2' = (z_{Kf} - z_{Kd}) \lambda_1 \sigma_k \theta$$

$$a_3' = (1-\lambda_1)$$

$$a_4 = f_1 \sigma_k \lambda_1 .$$

The price term included on the right hand consists of the border import price relative to the domestic capital goods price; the protection proxy used in the regressions so far is not incorporated. One additional exogenous variable is included compared to the reference estimation (FX_{Cap}). Identification of the distribution parameter *a* is not possible as the regression intercept now incorporates two constant terms, namely *a* and f_0. The expected sign of a_4 is negative (positive short run elasticity of substitution and a negative foreign exchange elasticity of the rate of protection).

The government reaction of tightening price and non-price measures is only one possible reaction to a foreign exchange shortfall. The parameter f_1 in equation 5.10 is only expected to be significantly negative for those countries where a government reaction to balance-of-payment difficulties consists mainly of raising trade barriers. Since a devaluation of the currency is another adjustment possibility due to its stimulating effect on exports and dampening influence on import demand, a problem might arise in the estimations. The change of the dollar price-index of domestically produced goods is partly due to a change in the exchange rate. If a foreign exchange shortfall leads mainly to a currency devaluation, multicollinearity between FX_{Cap} and (P'_{MCap}/P_{Kd}) can occur.

5.3.6.2. Measuring foreign exchange availability

We suppose that the variable measuring foreign exchange availability is represented by an index of real import capacity in terms of imported capital goods (1980 = 1). Nominal import capacity is defined as the exports of goods plus balance on services plus balance on government and private transfers plus net long-term capital inflows plus adjustments for errors and omissions (UNCTAD 1988a, p.12, fn a). This figure is deflated by the import price of capital goods, derived from the aggregation and deflation of trade data (see Appendix C).

Instituting tariff changes or implementing non-tariff barriers to imports is a bureaucratic procedure, so that the effect of a foreign exchange shortfall on the price of imported capital goods is likely to be spread over several years. A partial adjustment process (in addition to the

Table 5.10:
Time structure of the impact of foreign exchange availability on trade policies

Country	t	t-1	t-2
Argentina	.31	.39	.30
Chile	.01	.57	.42
Columbia	.02	.49	.49
Ecuador	.14	.55	.31
Peru	.02	.47	.51
Uruguay	.05	.73	.22

1 Figures are weights calculated from performing a regression including current and lagged values of the real import capacity. As for these six countries all regression coefficients were negative, the weights are percentage shares of each coefficient in the sum of all three coefficients.

one already included) cannot be incorporated in equation 5.10 since structural parameters would not be identifiable any more. Instead of determining a predefined lag structure (as is the case for, e.g., the Almon lag), we include the current, lagged, and twice-lagged values of the real import-capacity variable in a separate estimation of equation 5.11 designed to determine the lag structure of the foreign exchange influence. If the coefficients of the three variables all show the expected negative sign, weights for the foreign exchange variable are calculated as each coefficient's share in the sum of all three coefficients.[42] Such weights can be calculated for six of the twelve countries. Results of this exercise are presented in Table 5.10. It appears that the major reaction to a change in the foreign exchange variable occurs in the period (t-1); only Argentina shows a substantial share of adjustment in the same year as the varying real import capacity. The influence of protectionist policies takes longest in Peru. For those countries in the sample for which a time structure of the effect of the foreign exchange variable on protectionist policies could not be clearly determined, weights were set equal to .2 for the current value of real import capacity, .5 for the lagged one and .3 for the value lagged twice, which by and large approximates the 'average' lag structure observed for the six countries in Table 5.10.

42 The inclusion of three additional variables reduces the already low number of degress of freedom and causes multicollinearity problems. The results of these regressions are therefore used to construct a single foreign exchange variable.

5.3.6.3. Regression results

Regression results are presented in Table 5.11. The variables are defined as in the reference case with the exception of Brazil, for which the domestic capital stock is derived from expenditure data rather than production data. The parameter a_4 carries the expected sign for eleven of the twelve countries. Mexico represents the only contradiction to the supposedly negative relationship between the internal price of imported capital goods and the foreign exchange variable.[43] Furthermore, the coefficient of a_4 is at least significant at the 5% level with the correct sign for seven countries. This indicates that governments do react to foreign exchange shortages by lifting the protection rate for imported capital goods.

The alternative method of modelling the internal price of imported capital goods generates values for the short run elasticity of substitution more or less similar to the use of the protection proxy in the reference case. We do not observe a systematic variation of the estimates. Some countries now have higher short run elasticity estimates than in the base case (Mexico, the Philippines, Peru), while the other countries exhibit lower values. The general conclusion obtained from the various modifications so far is nevertheless sustained: Most countries are in a position to substitute for capital goods, but the capital stocks can be regarded as complements in the short run.

43 The real import capacity in terms of capital goods imported from the industrialized countries is only one possible measure of 'foreign exchange availability'. Marquez (1985) has tested the influence of different foreign exchange proxies on the completion time of investment projects in developing countries. He finds that his estimation results vary considerably with the chosen foreign exchange measures. The same can be observed in the Mexican case. If we use the level of real reserves (RES) (in terms of imported capital goods and weighted like FX_{Cap} above) as a measure of foreign exchange availability, the Mexican estimation becomes (standard deviations of the parameter estimates in brackets below):

$$\underset{}{\ln(k)} = \underset{(.1216)}{.1246} + \underset{(.0478)}{.1870}\ln(p) + \underset{(.1369)}{.8025}\ln(k_{t-1}) - \underset{(.0034)}{.0051}t - \underset{(.0298)}{.0848}\ln(RES)$$

$Dh = -.4781$, $R^2_{adj} = .9877$, $N = 15$.

k stands for the capital stock ratio and p for the relative price ratio of equation 5.11. The coefficient of the foreign exchange variable takes on both the anticipated negative sign and is significantly different from zero at the 5% level of significance. Further, no serial correlation is detected. An analysis of the influence of different foreign exchange variables on the internal price of imported intermediate and capital goods, however, goes beyond the scope of this study.

Table 5.11: Modification 4 (foreign exchange availability) estimations for capital goods (1972-1987)[1]

Country	Estimator	a_0'	a_1'	a_2'	a_3'	a_4	σ_k	f_1	R^2_{adj}	Serial	stability-test
Argentina	OLS	.3118 (.1898)	.2391*** (.0488)	-.0084* (.0042)	.8265*** (.0590)	-.1830*** (.0439)	1.378** (.5673)	-.7654** (2974)	.9773	D*h* = -1.285	.4272
Brazil[2]	IV,GLS	1.209*** (.3700)	.3620*** (.0660)	.0511*** (.0121)	.3607* (.1731)	-.3145** (.1330)	.5663*** (.1406)	-.8687* (.4460)	.9748	LM = 3.724* rho = -.2202	16.63***
Chile	OLS	.1836*** (.0569)	.2035*** (.0548)	-.0216*** (.0045)	.6937*** (.0436)	-.1329*** (.0393)	.6644*** (.1734)	-.6531* (.3314)	.9889	D*h* = 1.433	3.174*
Columbia	OLS	-.1265* (.0631)	.2623*** (.0387)	.0159*** (.0050)	.8499*** (.0473)	-.2535*** (.0669)	1.748*** (.5186)	-.9663** (.3505)	.9732	D*h* = .5050	.0602
Ecuador	OLS	-.7092** (.2949)	.0024 (.0656)	.0362** (.0133)	.5794*** (.1691)	-.0957** (.0441)	.0058 (.1555)	-39.18 (1058)	.9938	D*h* = .5519	.9127
Indonesia	OLS	-.8046 (.5070)	.2195 (.2884)	.0432 (.0285)	.6037** (.2107)	-.0105 (.1861)	.5539 (.7404)	-.0478 (.9078)	.9820	D*h* = -.1165	.4078
Malaysia	OLS	-.0407 (.1582)	.4104* (.2167)	-.0191 (.0223)	.6849*** (.1419)	-.0496 (.1636)	1.303** (.4944)	-.1208 (.4525)	.9782	D*h* = 1.492	.2159
Mexico[3]	IV, GLS	.2698 (.3362)	.1848* (.0925)	-.0198 (.0146)	.7982** (.3085)	.0839 (.0719)	.9157 (1.439)	.4538 (1.877)	.9327	D*h* = 2.229 rho = -.1713	1.701
Peru	OLS	.1973*** (.0317)	.2730*** (.0836)	-.0085* (.0044)	.7600*** (.0964)	-.1292** (.0486)	1.138** (.4472)	-.4734** (.2111)	.9126	D*h* = -.5224	1.816
Philippines	OLS	-.0436 (.0581)	.1276 (.0942)	-.0049 (.0081)	.7794*** (.1864)	-.0276 (.1338)	.5783 (.7102)	-.2165 (1.074)	.8275	D*h* = .1122	.6210
Uruguay	OLS	.0606 (.1222)	.2162** (.0782)	-.0003 (.0052)	.8171*** (.0721)	-.1461** (.0575)	1.182* (.5769)	-.6757 (.4578)	.9581	D*h* = .0043	1.070
Venezuela	OLS	-.2668*** (.0806)	.1283** (.0562)	.0258*** (.0062)	.3203 (.2075)	-.0213 (.0337)	.1888** (.0832)	-.1659 (.2978)	.9506	D*h* = 1.108	1.004

1 See text for a description of the variables included. Standard deviations of the parameter estimates are in brackets below. The estimates of the standard deviation of the long-run elasticity and f_1 are only asymptotically unbiased. The same holds for the estimates of standard deviations if the IV estimator is employed.

2 The capital stock derived from gross domestic investment data is used for Brazil as it turned out that the inclusion of the capital stock from capital goods production data resulted in a wrong sign for a_1 in the reference and modification 1 (depreciation rate 10%) estimations.

3 Estimation period 1972-1986.

***, **, * Estimate is different from zero (and the null-hoptheses of the stability- and LM-tests are rejected) at the 1%, 5%, and 10% level of significance.

Incorporating the combination of the structural parameters in equation 5.11, which is then estimated by non-linear least squares, produces estimates of the long run elasticity of substitution (σ_k) and the foreign exchange elasticity of the rate of protection (f_1), which are also shown in Table 5.11.

The direct estimation of the foreign exchange elasticity of protection (f_1) yields estimates significant at the 5% level for Argentina, Columbia, and Peru and at the 10% level for Brazil and Chile. All of these five estimates imply an inelastic response of the internal price of imported capital goods compared to a change in the real import capacity, this being lowest for Peru (-.473) and highest for Columbia (-.966). The estimate for Uruguay is in the same range though insignificant, while the value of f_1 for Ecuador is very high, but unreliable due to the large standard error attached to it. The lowest values of f_1 can be traced for Indonesia and Malaysia, indicating that trade barriers are varied to a low degree in response to the availability of foreign exchange.[44] For Indonesia, the coefficient of determination of the estimation is high (.982), but only the parameter of the lagged endogenous variable is significantly different from zero in the linear regression. This phenomenon could be caused by multicollinearity and might hint at a systematic variation of the foreign exchange variable with the relative price term. We already mentioned that this is the case if the primary means of tackling a balance of payments crisis is a currency devaluation and not the implementation of restrictive trade policies. Indonesia has confronted its balance of payments and external debt difficulties primarily by major devaluations (28% in 1983 and 31% in 1986). Additionally, trade distortions were reduced by replacing non-tariff barriers through tariffs and by dismantling several trade monopolies.[45]

The impact of the inclusion of a foreign exchange variable on the long run elasticity of substitution compared to the reference case is generally relatively low and not systematic. Six estimations return higher values, six lower ones. Most significant changes of the estimated long run elasticity of substitution compared to those in the reference case take place for Brazil and Chile. In both cases, an estimate close to half of the respective estimate including the protection proxy can be found.[46] In the Brazilian case, it is now significantly different from 1 which would imply that the two investment goods are complements even in the long run.

---

44 Compare Table 2.7.

45 See 'Indonesia slowly creeps out of its decade-long crisis', Financial Times January 24, 1989 and 'Indonesian growth outstrips expectations', Financial Times August 17, 1989. See also Bhattacharya and Linn (1988) and Woo and Nasution (1989).

46 As for Brazil domestic capital stock calculations are based on the expenditure concept, the estimate is compared to the one shown in Table 5.9.

Table 5.12: Spearman rank correlation coefficients between the estimated elasticities of substitution (capital goods)

parameter	$\sigma_R{}^1$	$\sigma_{.10}{}^2$	$\sigma_{GDI}{}^3$	$\sigma_{FX}{}^4$
$\sigma_R{}^1$	.	.9636***	.6000*	.8545***
$\sigma_{.10}{}^2$	.9636***	.	.5515*	.9091***
$\sigma_{GDI}{}^3$	.6000*	.5515*	.	.5878*
$\sigma_{FX}{}^4$	.8545***	.9091***	.5878*	.

Source:	Tables 5.6, 5.7, 5.9, 5.11 (compare Appendix A, Table A.1).
1	Calculated elasticities of substitution, reference estimations.
2	Calculated elasticities of substitution, modification 1 (depreciation rate 10%).
3	Calculated elasticities of substitution, modification 3 (gross domestic investment data).
4	Calculated elasticities of substitution, modification 4 (foreign exchange availability).
***	Spearman coefficient of rank correlation different from zero at 1% level of significance.
*	Spearman coefficient of rank correlation different from zero at 10% level of significance.

5.3.7. Comparison of estimated elasticities

Before analyzing the determinants of imported intermediate goods, the estimates of the elasticity of substitution between imported and domestically produced machinery and equipment are compared. If one excludes the user cost concept modification (as the necessary data was only available for five countries), four estimates of σ_k are obtained, which are termed σ_R (reference estimation), $\sigma_{.10}$ (depreciation rate 10%), σ_{GDI} (investment expenditure data), and σ_{FX} (foreign exchange). Spearman rank correlation coefficients are calculated between these four estimates, the results are given in Table 5.12 (compare Appendix A, Table A.1). If a long run elasticity could not be computed since the short run value carries a negative sign (Brazil for σ_R and $\sigma_{.10}$, Ecuador for σ_{GDI}), the value is not included in the rank correlation.

As can be traced from the table, the three estimates of σ_k based on the production statistics (σ_R, $\sigma_{.10}$, σ_{FX}) all exhibit rank correlation coefficients significant at the 1% level among themselves. Their positive correlation with the elasticity estimate of σ_{GDI} is lower. It is important to note this modification (gross domestic investment data) made use of both a different endogenous as well as an exogenous variable; nevertheless the estimates of it show a rank correlation coefficient significant at the 10% level with the three other elasticity estimates.

5.4. Imported intermediate goods

5.4.1. The stochastic equation

Analogous to the derivation of the estimating equation for capital goods, equation 4.24 can be written as

$$(5.12)\qquad \ln\left[\frac{Y_t}{M_{Int}}\right] = b_0 + b_1 \ln\left[\frac{P_{MInt}}{P_Y}\right] + b_2 t + b_3 \ln\left[\frac{Y_{t-1}}{M_{Int,t-1}}\right] + u_{12}$$

with

$$b_0 = \sigma_m \lambda_2 \ln\left[\frac{c}{1-c}\right]$$

$$b_1 = \sigma_m \lambda_2$$

$$b_2 = z_M \lambda_2 \sigma_m \tau$$

$$b_3 = (1-\lambda_2).$$

Y_t represents constant domestic value added of the economy in 1980 dollars, M_{Int} stands for real imports of intermediate goods in 1980 dollars, P_Y is the dollar price index of value added (1980 = 1), and P_{MInt} is a dollar price index of the internal price of imported intermediate imports (1980 = 1). The coefficient measuring the short run elasticity of substitution between imported intermediate imports and the bundle of domestically produced goods is b_1. It is the product of λ_2, the adjustment parameter, and the long run elasticity parameter σ_m. Finally, z_M measures the rate of disembodied technological progress of the imported intermediate products. The stochastic equation 5.12 allows for identification of all parameters appearing in the production function 4.8.

5.4.2. Estimation using the protection proxy

5.4.2.1. Data employed

In contrast to computations of the relationship between imported and domestically produced capital goods, all relevant variables of equation 5.12 can be observed directly or derived easily with the exception of the internal price of imported intermediate goods. No arbitrary assumptions have to be made with respect to such variables as the depreciation rate or user cost.[47] Therefore, we estimate only two different models; one uses the protection proxy, while the other introduces the foreign exchange variable to model protectionist policies.

Both the border price of intermediate imports and the import quantity measured in 1980 dollars are derived from deflating and aggregating trade data (Appendix C).[48] The border price is then corrected by multiplying with the country-specific protection proxy to derive an estimate of the internal price of imported intermediate products.

5.4.2.2. Estimation of the regression parameters

Estimates of the regression parameters are presented in Table 5.13. The overall fit of the regressions is much lower than that for the capital stocks. While the lagged endogenous variable played an important role in the latter estimations, its role in explaining the variance of the endogenous variable is now reduced for most countries; the coefficient b_3 even carries the wrong sign for Ecuador, Mexico, and Venezuela, but all three coefficients are insignificantly different from zero. Only Brazil exhibits a highly significant estimate of b_3.

The parameter b_1 carries the expected sign for all countries and it is not significantly different from zero (at the 10% level) only for Brazil, Indonesia, the Philippines, and Venezuela. Some estimates might be affected by inflated variances due to multicollinearity which will be explored below. Estimates of the short run elasticity of substitution between imported intermediate goods and domestic factors of production are highest for Columbia (.821), Ecuador (.628), Peru (.628), and Mexico (.604). While all four estimates are significantly different from zero, the null-hypothesis that they are equal to 1 cannot be rejected in either case at the 95% confidence level.

The F-test in Table 5.13 has the same meaning as in the previous tables. It indicates whether the inclusion of two slope-shifting dummy variables improves the explanation of the endogenous variable significantly. The null-hypothesis that at least one structural break exists (either in 1982 or 1985) is rejected for all countries but Brazil and Venezuela.

---

47 Due to lack of data, inventory changes cannot be registered.

48 A problem arises as not all countries report trade data until 1987. Intermediate imports therefore cover total imports of intermediate goods for some countries (Argentina, Brazil, Columbia, Malaysia, Uruguay), while they are restricted to imports from the industrialized countries for the others. Other intermediate imports, largely in energy-intermediates, are not covered. But the share of this product category in total imports is relatively low for most countries (compare Table 2.1 which reports the distribution of total imports).

Table 5.13: Estimation results for intermediate goods using the protection proxy, linear specification (1972-1987)[1]

Country	Estimator	b_0	b_1	b_2	b_3	R^2_{adj}	Serial	stability (F-test)
Argentina	OLS	2.671*** (.8288)	.2429* (.1361)	.0176* (.0093)	.1391 (.2572)	.4897	LM = 1.707	.5151
Brazil	OLS	.8182** (.3727)	.0321 (.1252)	.0227* (.0108)	.6171*** (.1794)	.9078	D*h* = -.1609	6.925**
Chile	OLS	1.861** (.6750)	.2822** (.1247)	.0057 (.0062)	.3794* (.2128)	.4338	D*h* = -.7131	.1266
Columbia	OLS	1.521** (.6384)	.8210*** (.2139)	.0017 (.0081)	.3716 (.2333)	.7964	D*h* = -1.614	1.276
Ecuador	OLS	2.992*** (.8541)	.6283*** (.1919)	.0223*** (.0068)	-.2124 (.3324)	.5196	LM = 3.375*	2.688
Indonesia	IV, GLS	2.060 (1.234)	.3274 (.2278)	.0157 (.0119)	.2656 (.4038)	.4264	LM = 7.906** rho = -.4883	.3668
Malaysia	IV, GLS	1.486** (.5326)	.5828** (.2412)	-.0125 (.0108)	.0246 (.3912)	.5212	D*h* = 2.588 rho = -.5052	.6005
Mexico	IV, GLS	3.409 (2.214)	.6044** (.2449)	-.0326 (.0201)	-.0206 (.6500)	.5093	D*h* = 3.857 rho = -.2992	1.339
Peru	IV, GLS	1.709 (1.663)	.6278** (.2777)	.0286 (.0320)	.3286 (.6650)	.6337	LM = 6.375** rho = -.3389	1.044
Philippines	OLS	2.346*** (.6736)	.0136 (.2265)	.0090 (.0053)	.0345 (.2824)	.1665	LM = 1.612	.4830
Uruguay	IV, GLS	2.051** (.7068)	.4395** (.1433)	.0096 (.0086)	.0605 (.3203)	.7282	LM = 5.362** rho = -.3129	.7243
Venezuela	OLS	2.993*** (.9501)	.3827 (.2908)	-.0113 (.0094)	-.0198 (.3216)	.1843	LM = 1.875	11.77***

1 See text for a description of the included variables. Standard deviations of the parameter estimates are in brackets. They are only asymptotically unbiased when the instrumental variable estimator is used.

***, **, * Estimate is different from zero (and the null-hypotheses of the stability and LM tests are rejected) at the 1%, 5%, and 10% level of significance.

5.4.2.3. Non-linear estimation of the structural parameters

We now incorporate the non-linear combination of the structural parameters in equation 5.12 and estimate it with non-linear least squares. Results are presented in Table 5.14. If the coefficient of the lagged endogenous variable is insignificantly different from zero, the variable is dropped, so that no difference is made between the short versus long run estimates of the elasticity of substitution.

The estimate of σ_m is significant at least at the 10% level for ten of the twelve countries. Corresponding to insignificant short run values, both Brazil and the Philippines possess non-significant estimates of σ_m. Expenditures on oil products account for a large share in the total import bill of Brazil, ranging up to 50% of total outlays. Although Brazil attempted to reduce its dependence on imported energy by instituting import-substitution programs in the 1970s and 1980s, the share of 'non-competitive imported inputs' (Fritsch 1988, p.256) in the total import bill remained high. "Brazil was especially vulnerable to the oil, interest rate, and debt shocks because of its longstanding emphasis on import substitution industrialization. There was no margin of substantial competitive imports to be reduced easily... In addition, Brazil remained very dependent on imported oil" (Cardoso and Fishlow 1989b, p.90). Other studies support the view that substitution possibilities for imported intermediates are apparently also limited in the Philippine economy (Dohner and Intal 1989, Rashid 1984).[49]

Columbia is the only country in the sample for which an estimate of the long run elasticity of substitution above unity appears in Table 5.14 and for which imported intermediate and domestically produced goods can be regarded as substitutes. In addition to Columbia, the results for Malaysia, Mexico, and Peru indicate that function 4.8 can be approximated by a Cobb-Douglas function, as the estimate of σ_m is not significantly different from 1.

Finally, the parameter estimates of z_M, the rate of technological progress, are low and have the expected positive sign for most countries. Positive and significant estimates can be observed for Argentina, Brazil, Ecuador, Peru, and the Philippines. Mexico, on the other hand, exhibits a negative and significant parameter estimate.

---

49 Compare also Zaidi (1988) who employs a fixed relationship between imported intermediate (and capital) goods and total output of the Philippine economy in a rationing model of imports. Nunnenkamp (1983) examines the impact of higher oil prices on manufacturing value added in seven newly industrializing countries (Brazil, Chile, Columbia, the Philippines, Singapore, South Korea, Tunisia). Interestingly, he finds a significantly negative impact of rising oil prices on industrial production for Brazil and the Philippines. These are the same countries which show non-significant short run elasticities of substitution for imported intermediate goods above.

Table 5.14: Estimation results for intermediate goods using the protection proxy, non-linear specification (1972-1987)[1]

Country	λ_2	σ_m	z_M	c	$R^2_{adj.}$	Serial
Argentina	1[2]	.2418* (.1236)	.0171** (.0060)	.9998*** (.0040)	.5175	DW=2.021
Brazil	.3829* (.1794)	.0838 (.3460)	.0547* (.0254)	.9981*** (.0160)	.9078	D*h*=-.1609
Chile	.6206** (.2128)	.4548* (.2473)	.0063 (.0071)	.9986*** (.0058)	.4338	D*h*=-.7131
Columbia	1[2]	1.049*** (.1686)	-.0036 (.0031)	.9179*** (.0321)	.7494	DW=1.725
Ecuador	1[2]	.5380*** (.1268)	.0135*** (.0034)	.9895*** (.0122)	.5017	DW=2.417
Indonesia	1[2]	.4000* (.2030)	.0092 (.0069)	.9835*** (.0125)	.3873	DW=1.053 rho=-.4553
Malaysia	1[2]	.5864** (.2134)	-.0078 (.0057)	.9272*** (.0552)	.5310	DW= .8431 rho=-.5051
Mexico	1[2]	.5753** (.2206)	-.0225*** (0055)	.9966*** (.0069)	.5366	DW=1.263 rho=-.3262
Peru	1[2]	.6296** (.2589)	.0265** (.0087)	.9834*** (.0276)	.6157	DW=1.226 rho=-.3389
Philippines	1[2]	.0753 (.2354)	.0099** (.0045)	.9998*** (.0001)	.1857	DW=1.719
Uruguay	1[2]	.4465*** (.1269)	.0070 (.0050)	.9926*** (.0101)	.7662	DW=1.168 rho=-.3546
Venezuela	1[2]	.3768* (.2104)	-.0081 (.0064)	.9975*** (.0017)	.2161	DW=1.914

1 See text for a description of the variables included. Asymptotic standard deviations of the parameter estimates are in brackets.

2 The speed of adjustment is restricted to one in the non-linear estimation as the estimate of b_3 in Table 5.13 is not significantly different from 0.

*** Parameter estimate is different from zero at the 1% level of significance.

** Parameter estimate is different from zero at the 5% level of significance.

* Parameter estimate is different from zero at the 10% level of significance.

5.4.3. Foreign exchange availability

We apply one modification to equation 4.24. The protection variable used to proxy the impact of tariff and non-tariff barriers on the border price of imported intermediate goods is replaced by a foreign exchange variable in the same way as this was done for capital goods in Section 5.3.6. The variable included in this case is the real import capacity in terms of imported intermediate goods. Nominal import capacity is defined in the same way as above; the figure is deflated by the dollar price index of imported intermediate goods.

5.4.3.1. The estimating equation

Corresponding to equations 5.8 and 5.9 above, the internal price of imported intermediate goods can be split into two components: the border price (P'_{MInt}) and the effect of protection ($1+t_{Int}$), which is again modelled as a function of real import capacity (FX_{Int}).

$$(5.13)\quad P_{MInt} = P'_{MInt}\,(1+t_{Int}),$$

$$(5.14)\quad (1+t_{Int}) = g_0\, FX_{Int}^{\,g_1}.$$

The foreign exchange elasticity of protection for intermediate imports is here represented by the parameter g_1 ($g_1 \leq 0$). Combining equations 4.24, 5.13, and 5.14, assuming a multiplicative error term of the exponential form, and taking logs on both sides, yields the estimating equation

$$(5.15)\quad \ln\left[\frac{Y_t}{M_{Int}}\right] = b'_0 + b'_1 \ln\left[\frac{P'_{MInt}}{P_Y}\right] + b'_2 t + b'_3 \ln\left[\frac{Y_{,t-1}}{M_{Int,t-1}}\right] + b_4 \ln(FX_{Int}) + u_{15}$$

with

$$b'_0 = \sigma_m \lambda_2 \left\{ \ln\left[\frac{c}{1-c}\right] + \ln(g_0) \right\}$$

$$b'_1 = \sigma_m \lambda_2$$

$$b'_2 = z_M \lambda_2 \sigma_m \tau$$

$$b_3' = (1-\lambda_2)$$

$$b_4 = g_1\sigma_m\lambda_2$$

The relative price term on the right hand side includes only the border price of imports. As was the case when examining the relationship between real import capacity and the internal price of capital goods, the distribution parameter *c* cannot be recovered from the regression coefficients as the rate of protection on intermediate goods invariant to the level of available foreign exchange (g_0) enters the intercept term as well. The foreign exchange variable is presumed to be negatively correlated with price-raising protectionist actions of the government ($b_4 \leq 0$).

5.4.3.2. Regression results

Results of estimating equation 5.15 with ordinary and non-linear least squares (substituting the non-linear parameter combinations) are shown in Table 5.15. We again exclude the lagged endogenous variable in those cases where its regression coefficient turned out negative or insignificantly different from zero in the initial estimation.

As was the case when examining the relationship between the foreign exchange variable and protectionist measures levied on imported capital goods, the sign of b_4 for Mexico carries an unexpected positive sign. The additional explanatory power of the included variable is low for most countries. Only for Columbia and Malaysia is the parameter of the import capacity index significant at the 10% level. Several previously significant estimates of the short run elasticity of substitution become insignificant (Chile, Indonesia, Venezuela), which might be due to the presence of multicollinearity. Otherwise estimates of b_1' (which is equal to σ_m for all countries except Brazil and Chile) are close to the calculated coefficients obtained when we employed the protection proxy.

The direct estimate of g_1, the foreign exchange elasticity of the protection rate levied on imported intermediate goods, is only significant for Peru, where it takes a slightly lower magnitude as the estimate of the same parameter when considering capital goods (-.398 vs. -.473). Most other estimates of g_1 are insignificant and relatively low (Argentina, Chile, Columbia, Ecuador, and Uruguay); they are at least smaller than the estimate of the foreign exchange elasticity of protection on capital imports when this parameter was computed. Due to very high standard deviations not much can be said about the magnitude of g_1 for Brazil, the Philippines, and Venezuela. These results indicate a much weaker link between foreign exchange availability and the protectionist measures in the case of imported intermediate goods than of imported capital goods.

Table 5.15: Modification estimations (foreign exchange availability) for imported intermediate goods (1972-1987)[1]

Country	Estimator	b_0'	b_1'	b_2'	b_3'	b_4	g_1[3]	R^2_{adj}	Serial
Argentina	OLS	3.592*** (.1031)	.3034* (.1564)	-.0313*** (.0068)	0[2]	-.1414 (.1090)	-.4661 (.5013)	.6149	DW=2.001
Brazil	OLS	.7501** (.3421)	.0694 (.1137)	.0217* (.0103)	.6336*** (.1713)	-.1654 (.1312)	-2.385 (4.217)	.9021	D*h*=.0682
Chile	OLS	1.888** (.8434)	.1616 (.1685)	.0018 (.0119)	.3986 (.2933)	-.0170 (.1551)	-.1053 (1.029)	.1662	LM=.0109
Columbia	OLS	2.387*** (.1970)	.9075*** (.1794)	.0129 (.0141)	0[2]	-.4108* (.2282)	-.4527 (.2937)	.7126	DW=1.433
Ecuador	OLS	2.423*** (.1330)	.5024*** (.1373)	.0238** (.0098)	0[2]	-.0830 (.1203)	-.1651 (.2467)	.4794	DW=2.363
Indonesia	GLS	2.924*** (.2922)	.3127 (.4524)	-.0172 (.0317)	0[2]	-.0653 (.3339)	-.2093 (1.286)	.3486	DW=1.149 rho=-.3994
Malaysia	GLS	1.089*** (.2792)	.6032*** (.1775)	.0273 (.0230)	0[2]	-.4909* (.2471)	-.8139 (4930)	.6938	DW=1.135 rho=-.3509
Mexico	OLS	3.574*** (.1646)	.6956** (.2449)	-.0488** (.0163)	0[2]	.1386 (.1524)	.1992 (.1942)	.5574	DW=1.421
Peru	GLS	2.761*** (.1639)	.7998** (.3123)	.0236 (.0137)	0[2]	-.3180 (.1934)	-.3976** (.1698)	.6919	DW=1.231 rho=-.3570
Philippines	OLS	2.318*** (.1724)	.2241 (.2631)	.0170 (.0143)	0[2]	-.1023 (.3085)	-.4568 (1.311)	.4506	DW=1.650
Uruguay	GLS	2.278*** (.1358)	.2621* (.1406)	.0089 (.0103)	0[2]	-.0064 (.1536)	-.0244 (.5625)	.5340	DW=1.312 rho=.3237
Venezuela	GLS	2.814*** (.1264)	.0850 (.2597)	.0040 (.0107)	0[2]	-.3295* (.1707)	-3.410 (7.813)	.3353	DW=2.621 rho=.2683

1 See text for a description of the variables included. Standard errors of the parameter estimates are in brackets.
2 The parameter of the lagged endogenous variable is restricted to 0 as it is not significantly different from zero.
3 Asymptotic standard error only.
***, **, * Parameter estimate is different from zero at the 1%, 5%, 10% level of significance.

5.5. Summary and interpretation of the empirical results

5.5.1. Summary of the empirical results

This chapter examined the relationship between imported and domestic factors of production in twelve developing countries. Rather than determining specific values of substitution elasticities, the purpose of the chapter was primarily to answer the broad question of whether domestically produced and imported capital and intermediate goods have to be described as complements or substitutes in an aggregate macroeconomic production context.

We estimated short and long run elasticities of substitution assuming that individual firms determine their input choice in an optimal way. Since data on several of the necessary variables are not available, assumptions had to be made in order to compute both price and quantity variables. In order to examine whether the results of the econometric exercises are sensitive to these premises, we varied several assumptions about data formation and compared the results. The most important modification concerned the inclusion of a behavioral equation, which models government trade policy as a function of foreign exchange availability. Estimation of this modified version implicitly computed the foreign exchange elasticity of the protection rate for imported capital and intermediate goods. We can draw the following conclusions from the results:

(a) Imported capital goods are complements to domestically produced investment goods in the short run, although a fixed-proportions approach appears to be justified for only two countries, namely Ecuador and the Philippines. The reference estimation and modifications thereof have shown that changes in the relative user cost of the two capital products affect the composition of the efficiency capital stock even in the short run.

(b) The largest alteration of the results for the estimated elasticity of substitution between internally produced investment goods and those manufactured in the industrialized countries resulted from using expenditure instead of production data as the basis for computing the domestic capital stock. Estimated elasticities were generally found higher when production data were employed. Expenditure data on gross domestic investment include a much higher share of government investment, for example on infrastructural projects. This result could therefore be expected, since the bundle of domestically produced capital goods comprises non-tradable services such as construction of buildings and streets.

(c) The adjustment of the capital stocks to their desired level after a price change takes considerable time due to delivery lags and limited downward adjustment possibilities (the mean lag is mostly between 2 and 3 years). Correspondingly, the short and long run estimates of the elasticity of substitution between domestically produced and imported investment goods differ substantially. It appears that the long run relationship between the two capital stocks can be described in a Cobb-Douglas functional form ($\sigma_k = 1$) for Argentina, Chile, Columbia, Malaysia, Peru, and Uruguay.[50] Estimates of σ_k for Indonesia, Mexico, and Venezuela were found to be significantly different from zero, but also significantly different from 1 in at least one computation. The two capital goods can be described as complements for these countries. Finally, no or only weak evidence has been gathered that would imply a non-zero elasticity of substitution for Ecuador and the Philippines. When modelling these economies in a macroeconomic framework, a fixed-proportions production function seems to be appropriate.[51]

The Brazilian estimates represent a special case. Calculations based on the derivation of the domestic capital stock from data on domestic capital goods production were unsatisfactory. On the other hand, if we employed expenditure data, σ_k not only assumed the correct sign but was significant both in the short and long run. While the calculated long run estimate was slightly above 1 (characterizing the two capital stocks as substitutes) using the protection proxy, it fell below 1 when the effect of tariff- and non-tariff measures was described as a function of foreign exchange availability. It is possible that errors in variables have caused the unsatisfactory results when production data were employed to estimate the domestic capital stock K_d.

(d) While adaptation of the capital stocks to their long run levels takes considerable time, adjustment lags played a minor role in the estimations for imported intermediate goods. Significant lags were found only for Brazil and Chile. Correspondingly, the computed short and long run elasticities of substitution between domestic means of production and imported intermediate ones do not differ significantly for most countries.

(e) Using the protection proxy in order to describe the internal price of intermediate goods, all countries but Brazil and the Philippines exhibit significant elasticities of substitution for imported intermediate goods. Estimates in Table 5.14 indicate that they are complements to domestic value added for most countries. It is not possible to reject the null-hypothesis that the elasticity is equal to 1 (and that the resulting functional form of equation 4.8 is of a Cobb-Douglas nature) for Columbia, Malaysia, Mexico, and Peru. These results support other studies which describe domestic means of production and imported inputs as complements in developing countries. Marquez (1984) and Marquez and Pauly (1987) estimate the substitution elasticity between the capital stock and imported oil in non-oil developing

---

50 The Argentinian estimate of the elasticity of substitution between imported and domestically manufactured capital stocks (average around 1.5) is somewhat lower than the estimate by Michalopoulos (1975), who calculated a value of 2.3 for the period 1949-1965 (using Argentinian Central Bank data), but in both cases a substitutability relationship characterizes the two stocks. One might judge that the inward-oriented policies did not result in an increased ability to substitute for imported machinery and equipment.

51 Compare Lucas (1989, p.145).

Figure 5.1:
Estimated elasticities of substitution for imported capital and intermediate goods[1]

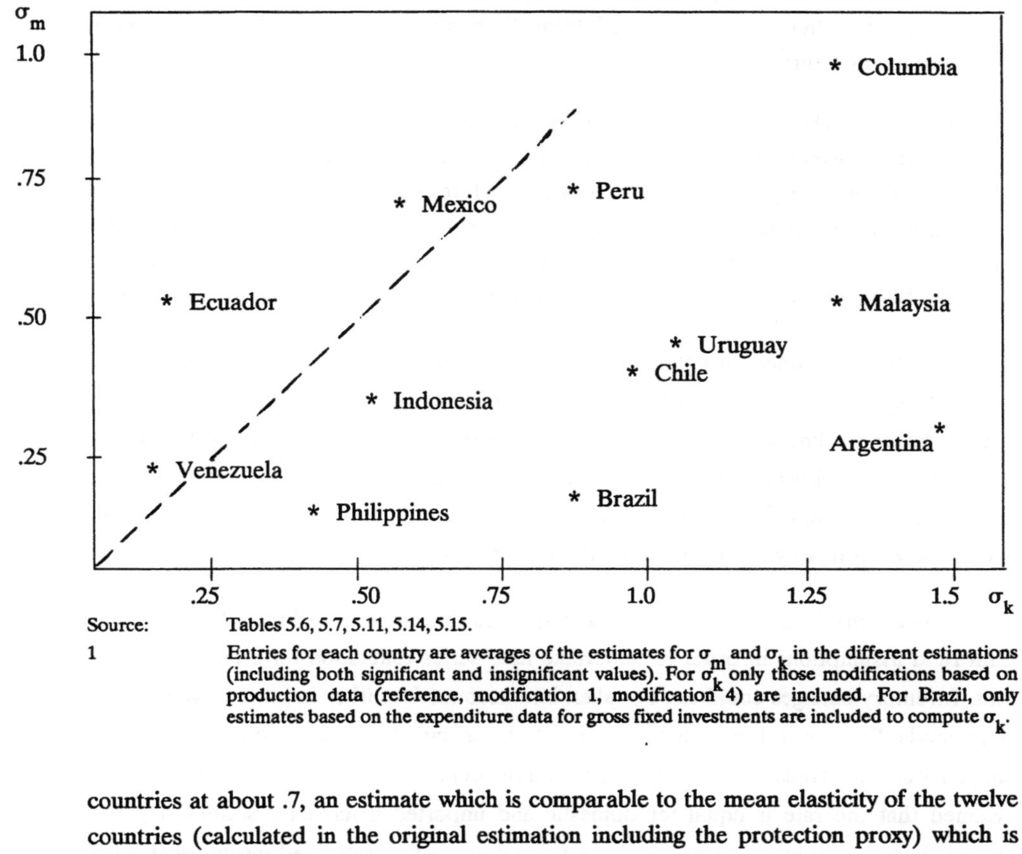

Source: Tables 5.6, 5.7, 5.11, 5.14, 5.15.

1 Entries for each country are averages of the estimates for σ_m and σ_k in the different estimations (including both significant and insignificant values). For σ_k only those modifications based on production data (reference, modification 1, modification 4) are included. For Brazil, only estimates based on the expenditure data for gross fixed investments are included to compute σ_k.

countries at about .7, an estimate which is comparable to the mean elasticity of the twelve countries (calculated in the original estimation including the protection proxy) which is somewhat below .5.[52]

(f) If the computed values between the two long run elasticities of substitution for the different import categories are compared, it emerges that the substitution possibility for imported capital goods is higher than for imported intermediate goods for most countries as depicted in Figure 5.1.[53] It appears that only in Ecuador (an oil-exporting country) is the reverse the case, with Venezuela and Mexico exhibiting only small differences between the two elasticity estimates. The finding is broadly in accordance with the general assessment of

52 The results might also be interpreted to support Berndt and Wood (1975) and Pindyck and Rotemberg (1983a), who find that capital and energy are complements. This assessment needs to be validated by treating energy imports as a factor of production separate from other intermediate imports. This was not possible here due to data limitations (see Appendix C).

53 For capital goods, only the reference case and modifications based on the production statistics are considered.

studies examining import price responses which find that price elasticities are higher for manufactures than for primary products (Goldstein and Khan 1985, p.1070), although one should note that our measure of intermediate imports comprises both manufactures and non-manufactures.

(g) We tested whether tariff and non-tariff measures used by the governments of the twelve countries respond to changes in real import capacity. The empirical evidence presented in this chapter suggests that such a functional relationship exists, but that governments discriminate between different import categories. While a relatively strong negative link between the foreign exchange variable and the leverage of protectionist measures could be established for capital goods, this could not be manifested for imported intermediates.

The findings concerning the magnitude of the long run substitution elasticities and the leverage of import restrictions are coherent. Since most countries possess a lower long run elasticity of substitution for imported intermediate than imported capital goods, foreign exchange resources are primarily directed towards intermediates if a balance of payments crisis occurs. In those economies where import restrictions are at least partly relied on to curb import demand, these policies target primarily capital and not intermediate goods.

Some remarks should be made with respect to the variables included. We had to make several rigid assumptions in order to undertake time series estimation for a broad range of countries. Inventory changes could not be taken into account for either imported capital or intermediate goods. Further, a depreciation rate had to be specified as country specific data for this parameter were not available. Although we varied the depreciation rate in one modification, we still assumed that the rate is equal for domestic and imported machinery. Studies that reexamine import elasticities using country specific data on depreciation rates and inventory changes would be valuable to cross-check some of the results obtained in this chapter.

It should also be kept in mind that we interprete the elasticities of substitution as *average values*. As we noted in Chapter 4, the estimated elasticities include the change of the output mix of the economy in response to a factor price change as well as factor substitution on the plant level relating to both new and already installed investments.

5.5.2. Interpretation of the empirical results

The spread of estimated long run substitution elasticities is much larger for imported capital than for intermediate goods, which suggests that some uniform influences exist that characterize imported intermediate goods and domestic means of production as complements in most countries. As was outlined in Chapter 2, non-traded factors of production are one of the major factors causing an imperfect substitutability for intermediate imports. The relatively uniform pattern of estimated elasticities hints at the important role these non-traded factors play in determining the dependence on imported intermediates.

One possible explanation for the observed divergence among the elasticity estimates between domestically manufactured and imported capital goods could stem from the different policies concerning the development of a capital goods sector. It was shown in Chapter 2 that some of the developing countries have pursued an active import substitution policy for quite some time, thereby reducing the contribution of imported machinery to apparent consumption of capital goods. On the other hand, it has also been observed that some countries strengthened the capital goods sector without reducing the share of foreign made equipment but seeking export possibilities of their own.

Table 5.16 gives Spearman rank correlation coefficients of the estimated elasticities of substitution between imported and domestically manufactured capital goods (based on the production statistics) and several indicators of the capital goods sector used in Chapter 2. The first is the range of activities in the capital goods sector (in 1980) and one might expect a positive correlation to the substitution elasticities. The second is the average share of imported capital goods in apparent consumption of capital goods (1972-1987). We observed in Chapter 2 that several large developed countries report far higher shares of this figure than several developing countries studied here. The contribution of imported capital goods to the total supply of capital goods is thus not merely a function of size and income but also of the development strategy. The main motivation for moving into secondary import substitution for developing countries rests with an alleged gain in 'independence', if imports of previously imported machinery and equipment are curtailed. According to this view, a negative correlation is expected. Finally, the third indicator included in the rank correlation analysis is the average share of GDP originating in capital goods industries (1972-1987). It measures the strength of the capital goods industries in each country. A positive correlation is anticipated.

The result of this analysis is striking. Only a weak correlation between the estimated elasticities and the range of activities in the capital goods sector can be traced from the table, which shows an unexpected negative sign in two instances. It appears that a broad range of different manufacturing activities in the capital goods sector does not guarantee high substitution elasticities.[54]

---

54 It should be noted that the number of observations for the rank correlations including the 'range of activities' is relatively small as data for this variable are missing for Ecuador and Uruguay. Additionally, σ_k could not be calculated for Brazil in the reference case and modification 1. See Appendix A, Table A.1 for the data included.

Table 5.16:
Spearman rank correlation coefficients between estimated elasticities of substitution (capital goods) and selected indicators

Parameter	Range of activities[1]	Average share of imported capital goods in apparent consumption[2] (1972-1987)	Average share of GDP originating in capital goods industries[3] (1972-1987)
σ_R[4]	-.2333	-.4636	.6455**
$\sigma_{.10}$[5]	-.1500	-.4273	.6273**
σ_{FX}[6]	.1833	-.5182	.6000*

Source: Calculations based on data presented in Appendix A, Table A.1.
1 Number of activities in the capital goods industries. See Appendix A, Table A.1.
2 Average share of imports of capital goods in apparent consumption of capital goods (production minus exports plus imports), 1972-1987 (for Mexico 1972-1986). For data see Appendix A, Table A.1. The definition of capital goods production and trade follows Appendix C, Table C.1.
3 See Table A.1 in Appendix A for the data.
4 Calculated elasticities of substitution, reference estimations.
5 Calculated elasticities of substitution, modification 1 (depreciation rate 10%).
6 Calculated elasticity of substitution, modification 4 (foreign exchange availability).
** Spearman coefficient of rank correlation different from zero at 5% level of significance.
* Spearman coefficient of rank correlation different from zero at 10% level of significance.

Although the rank correlation between the share of imports in apparent consumption and the elasticity estimates carries the expected negative sign, none of the coefficients is significant. Conversely, the measure of the strength of the capital goods sector is significantly positively correlated with the estimated substitution elasticities; twice at the 5% level of significance and once at the 10% level.

The finding has important implications since it touches on the broad questions of development strategies. No empirical support can be found for the view that a curtailment of imported capital goods raises the elasticity of substitution and thereby reduces the dependence on imported machinery. This implies that the obvious disadvantages stemming from such a strategy (like the inefficient distribution of resources) are not counterbalanced by a reduction of the reliance on imported investment goods. One should recall at this point that capital goods manufacturing is characterized by economies of specialization which have led to increased international intra-industry trade in machinery and equipment. Squeezing imports of capital goods thereby implies that an economy encounters the danger of being detached from international technological progress with strong implications for domestic research and development efforts, productivity growth etc. Further, the overall reliance of the economy on imported machinery is not reduced as the suppressed quantity of imported capital goods remains important for the domestic industries. The negative impact of a foreign exchange shortage is not alleviated by fostering domestic capital goods industries through suppressing imports.[55]

On the other hand, the strength of the capital goods sector is an important determinant of substitution possibilities for imported capital goods as skills and indigenous technological know-how are acquired especially in industrial activities that manufacture machines and equipment.[56] But as is stressed by Jacobsson (1988), it is important that this development does not take place in a highly protected environment which creates internationally uncompetitive industries. Rather, openness to trade is especially important in this branch, as trade in capital goods is one (if not the most) important medium of technology transfer.

---

55 In order to crosscheck this assessment it would be valuable to derive a price-indicator which includes all measures that affect the relative user costs of the two capital goods and calculate the rank correlation between the elasticities of substitution and this indicator. It would then also be possible to judge whether more outward-oriented development strategies (an overall non-distorted price ratio) do indeed lead to higher substitution elasticities.

56 Michalopoulos (1975, p.353) and Roemer (1981, p.429) emphasize that the strength of the capital goods industries is an important determinant of the degree of technological independence and of substitution possibilities for imported capital goods.

6. Imports, growth, and the trade balance: two case studies

6.1. Basic remarks about the framework

Faced with tightening international capital markets, all highly indebted developing countries experienced a major activation of their trade balance positions in the eighties. While in some countries the trade balance improvement was accompanied by decreasing GDP growth rates, others were able to maintain pre-crisis growth rates with an accomplished external account amelioration by relocating production towards the sector producing tradable goods. This chapter explores how the different structural importance of imported intermediate and capital goods that was traced in the preceeding empirical analysis influences the reaction of the trade balance to changes in key relative prices, the domestic growth rate, and world income. Thus, it is studied to what degree the import-intensity of domestic production affects the latitude to pursue external adjustment policies. Case studies are conducted for two countries which were found to be characterized by different production structures: one is Columbia - able to substitute for both types of imported factors of production to a relatively high degree - and the other is Ecuador, where substitution capabilities have turned out to be limited.

We investigate the linkages between merchandise trade and domestic growth within a partial equilibrium framework. Specifically, focus rests on the supply-side of the economy under consideration while aggregate demand and its determinants are not explored.[1] The framework we present is similar in structure to the import compression model by Khan and Knight (1986, 1988). They specify an export-supply function of the Cobb-Douglas form with imported inputs and value added by domestic factors of production as arguments. In addition to modelling world demand for exports, they present an import demand function which incorporates the stock of international reserves as a measure for foreign exchange stringency besides a relative price term. The underlying hypothesis is that the government compresses imports whenever a need to accumulate

1 Fritsch (1988) argues that the linkage between imports and domestic supply is of primary importance; aggregate production corresponds more to 'capacity' or 'potential' output in such frameworks rather than to actual output levels. Alternatively, one could envisage that domestic demand management policies target total demand to supply capacities (Khan and Knight 1986, p.17). Development and debt-cum-growth models that build on the two-gap approach of Chenery and Strout (1966) stress this supply linkage. See Glick and Kharas (1986) and Hentschel (1988) for surveys. Import compression models generally operate in the same partial equilibrium framework (compare Currie et al. (1988, p.13) for a short description of these models). Further, the model developed here concentrates on the real flow of goods and leaves aside the financial transactions going hand in hand with merchandise trade; the behavior of capital account components (like flows of short and long term capital or foreign direct investment) are not analyzed.

foreign exchange reserves emerges. Within this structural framework, Khan and Knight (1986) introduce several (exogenous) shocks and trace the dynamic paths of exports, imports, and the trade balance.

While Khan and Knight are mainly concerned with contrasting the evolution of these variables in their extended model to the results of the 'standard' import and export models (which do not take into account the direct production linkage of imported inputs), the model we present here focuses on the importance of different structural relationships between imports and productive capacity. Can trade balance improvements (measured in foreign currency), which are a precondition for the repayment of external debt, be achieved by a devaluation? Or are policies which reduce output growth and consequently demand for imported means of production necessary? Similarly, to what extent are the two countries able to absorb import price shocks?

It might be useful to sketch briefly how the trade flows are determined in the model. A one-sector economy is assumed. The system of production functions is in accord with the three level nested constant elasticity of substitution function described and estimated in Chapters 4 and 5. We specify derived factor demands for imported intermediate goods and the two capital stocks and estimate them under the assumption that firms minimize production costs subject to a given output level. The cost functions corresponding to the production functions included are deduced and the prices of domestic value added and the overall price level of gross output are endogenized. In this respect, the model presented here differs from that of Khan and Knight (1986, 1988), who assume domestic prices to be exogenous. Our approach allows us to observe how changing prices of imported inputs influence the domestic price level of value added and gross output and thereby undermine nominal exchange rate devaluations.

We assume both Columbia and Ecuador to be small on the import side so that import prices are given. This is a reasonable assumption since the countries import a wide variety of products and their respective market shares in total world imports of the individual products are small. Imports of intermediate goods are directly determined from the derived factor demand function; imports of capital goods are deduced from the change in the stock of imported machinery, taking into account depreciation on past investment. The residual import category ('other imports') consists mainly of consumer goods and constitutes a minor share in total imports. Demand for these goods is modelled as a function of domestic income and the relative price between this class of imports and the home-made goods.

We assume product differentiation on the export side. This allows us to incorporate an export supply function explicitly while staying within the framework of a one-sector economy (Melo and Robinson 1989). In contrast to the import side, export prices are not assumed to be determined exogenously. "While the assumption of an infinite price elasticity seems reasonable a priori in the case of the world supply of imports to a single country, this assumption carries far less intuitive appeal when applied to the supply of exports of an individual country. That is, unless idle capacity exists in the export (or domestic) sector, or more generally, unless export production is subject to constant or increasing returns to scale, it is unlikely that an increase in the world demand for a country's exports can be satisfied without any increase in the price of its exports (at least in the short run)" (Goldstein and Khan 1978, p.275). Columbia and Ecuador exhibit highly concentrated exports in a few commodities which account for more or less significant market shares in total world trade of these products. The export price is therefore formed endogenously by the interaction of supply and demand.

After estimating the parameters of the structural equations, it will be possible to analyze the impact of hypothetical developments of the exogenous variables (such as the exchange rate, import prices, or the domestic and world income growth rates) on the trade balance. Hence, emphasis lies on implied *changes* in the simulated trade balances rather than on the latters' absolute values; the trade account calculations are not at all intended to be 'forecasts', which are a feature of debt-projection models. The change in the total amount of debt is linked to computations of the trade and current accounts in such models. They incorporate assumptions about the probable future evolution of exogenous variables and calculate trade and current account balances from estimated export and import functions.[2] The simple model used here is structurally close to the debt-projection frameworks, but the link between the balances on the trade and current accounts and debt developments is not made. First, this is due to the above mentioned purpose which concentrates on evaluating the relative effect of policy and price changes on the trade balance rather than forecasting it. Second, the link between the trade and current account and debt developments requires projections of capital account items which has proven to be a very difficult task, often yielding unprecise results.

The next section introduces the Columbian model, which is first specified econometrically. After deriving each estimating equation, results of the computations are presented in order to ease discussion of the variables and parameters included. The model is then validated by ex post simulation. Exogenous shocks are introduced and the simulated paths of exports, imports, and the trade balance are presented. The third section deals with these same issues for Ecuador, which differs from the Columbian model with respect to the specification of the system of production functions.

---

2 Debt-projection models were very popular after the outbreak of the debt crisis. Compare Cline (1983, 1984), Dooley et al. (1986), Leven and Roberts (1983), Inter-American Development Bank (1984), Richtering (1989), and Saunders and Dean (1986). See Suttle (1986) for a survey of these models.

6.2. The Columbian model

6.2.1. Econometric specification

The model is specified and estimated in the national currency, pesos. All quantities are expressed in 1980 constant pesos (comprising imports and exports as well). The trade balance is expressed in current dollars as it is the relevant currency for the repayment of debt. The exchange rate is included explicitly in the equations in order to convert import prices into local currency.

6.2.1.1. Factor demands, price equations, and the demand for other imports

As in the previous chapters, we assume a three-level nested production function. Gross output (Q) of the economy is a function of domestic value added (Y), or GDP, and imported intermediate inputs (M_{Int}). The former is produced by the efficiency capital stock (J) and labor (L). Finally, the efficiency capital stock is a function of the stock produced from domestic investment goods (K_d) and the one made up of imported capital goods (K_f). Results of estimating the elasticity of substitution between value added and imported intermediate products, as well as between the two capital stocks, show that these are insignificantly different from 1 (compare Chapter 5). The two are therefore modelled as Cobb-Douglas functions.[3] The function relating J and L is also assumed to approximate the Cobb-Douglas form in the long run.[4] The three functions are:

$$(6.1) \qquad Q = \left[e^{(z_Y t)} Y \right]^{c} \left[e^{(z_M t)} M_{Int} \right]^{1-c}$$

$$(6.2) \qquad Y = J^{b} L^{(1-b)}$$

3 The hyper function is assumed to be linearly homogeneous of degree 1.

4 This hypothesis about the elasticity between labor and the efficiency capital stock cannot be tested since data on average man-hour employment could not be obtained on an economy-wide basis. Studies that examine the functional relationship between capital and labor in developing countries generally find substantial substitution possibilities in most industries, often close to unity. Compare Fitchett (1976), Hoffmann and Weber (1976), Katz (1969), Laumas and Williams (1981), Mulate (1980), Sankar (1970), and Sines (1979). See Morawetz (1976) for a survey and evaluation of earlier studies.

$$(6.3) \qquad J = \left[e^{(z_{Kd}t)} K_d\right]^{a} \left[e^{(z_{Kf}t)} K_f\right]^{1-a},$$

where *a*, *b*, and *c* are factor shares and e^{zi} measures the factor specific disembodied technological progress in K_d, K_f, Y, and M_{Int}.[5] Assuming cost-minimizing firms and combining the first-order conditions for a minimum for equation 6.1 gives[6]

$$(6.4) \qquad M^{*}_{Int} = \left[\frac{1-c}{c}\right] \left[\frac{P_Y}{P'_{MInt}\ E\ (1+t_{Int})}\right] Y,$$

where t_{Int} represents the tariff rate (calculated as described in Section 5.2.1. and Appendix C.4.1.), E is an exchange rate index of pesos per dollar (1980 = 1), and a star indicates the desired (long run) value of M_{Int}. P_Y and P'_{MInt} stand for the GDP deflator and the US-dollar index of the border price of imported intermediate goods (1980 = 1). Assuming an adjustment of actual to desired levels of the Koyck-lag form in equation 6.4 gives the following estimating equation:

$$(6.5) \qquad \ln(M_{Int}) = \lambda_{MInt} \ln\left[\frac{1-c}{c}\right] + \lambda_{MInt} \ln\left[\frac{P_Y}{P'_{MInt}\ E\ (1+t_{Int})}\right]$$

$$+ \lambda_{MInt} \ln(Y) + (1-\lambda_{MInt}) \ln\left[M_{Int,t-1}\right] + u_5,$$

with λ_{MInt} as the adjustment parameter. We assume that output prices are formed with a constant mark-up over production costs:

$$(6.6) \qquad P_Q = t_{PQ}\ (1+r) \left\{ P_Y \left[\frac{Y}{Q}\right] + P'_{MInt}\ E\ (1+t_{Int}) \left[\frac{M_{Int}}{Q}\right]\right\},$$

where r denotes the mark-up parameter and t_{PQ} is an indexation effect due to the inclusion of the tariff on intermediate imports..

---

5 Due to an identification problem, Hicks-neutral progress is assumed for the domestic value added function (Y). It can be picked up by including a technological progress term attached to Y at the hyper level.

6 Compare Section 4.3.4.

Substituting the derived factor demands for Y and M_{Int} from the production function into equation 6.6 gives the cost function for the long run desired value of P_Q:[7]

$$(6.7)\quad P_Q^{*} = t_{PQ}(1+r)\, P_Y^{\,c} \left[P'_{MInt}(1+t_{Int})E\right]^{1-c} \left\{\left[\frac{c}{1-c}\right]^{1-c} + \left[\frac{1-c}{c}\right]^{c}\right\} e^{-z_Q t},$$

with $\quad z_Q = z_Y c + z_M(1-c)$

as a term consisting of the technological progress of value added and labor, weighted by their respective factor shares in gross output. We use an adjustment process which links the alteration in P_Q to desired and past changes in prices (König 1990, p.208). This yields the following estimating equation:

$$(6.8)\quad \ln(P_Q) = P_{Q0} + \lambda_{PQ}\, c \ln(P_Y) + \lambda_{PQ}\,(1-c) \ln\left[P'_{MInt}\; E\;(1+t_{Int})\right]$$

$$- \lambda_{PQ} z_Q t + (1-\lambda_{PQ}) \left[2 \ln(P_{Q,t-1}) - \ln(P_{Q,t-2})\right] + u_8,$$

with

$$P_{Q0} = \lambda_{PQ} \ln\left[t_{PQ}(1+r)\right] + \lambda_{PQ} \ln\left\{\left[\frac{c}{1-c}\right]^{1-c} + \left[\frac{1-c}{c}\right]^{c}\right\}.$$

The factor demand for imported intermediate imports (equation 6.5) and the corresponding determination of the output deflator (equation 6.8) are estimated jointly with non-linear least squares. Regression results are shown in Table 6.1.[8]

The distribution parameter c in the Cobb-Douglas formulation stands for the long run share of domestic value added in nominal gross output, which is estimated at around 92%. Vice versa, the value of imported intermediate imports accounts for 8% of the nominal value of gross

---

7 Compare König (1990, Chapter 5) for specification and estimation of a system of Cobb-Douglas functions and their cost function counterparts.

8 Although estimating non-linear equations, we report the Durbin's h-statistic (and later also the Durbin-Watson statistic if no lagged endogenous variable is present in the estimating equation), as it is a frequently used measure for first-order autocorrelation even when non-linear estimations are conducted. Compare, for example, Conway (1986), Feltenstein et al. (1990), and König (1990).

Table 6.1:
Columbia: Joint NLLS estimation of the factor demand for intermediate imports (M_{Int}) and the output price equation (P_Q), 1972-1987

endogenous variable	$R^2_{adj.}$	serial	SSE
$\ln(M_{Int})$	.920	$Dh = -.0634$	.1024
$\ln(P_Q)$	.999	$Dh = -.2966$	.0160

parameter	estimate	stand. error	t-statistic
c	.9216	.0025	374.2
P_{Q0}	.0224	.0223	1.003
z_Q	.0059	.0039	1.512
λ_{PQ}	.5772	.1313	4.396
λ_{MInt}	.6976	.1460	4.775

output. The mean lag of the quantity of intermediate imports $[(1-\lambda_{MInt})/\lambda_{MInt}]$ is estimated to be less than half a year. Instantaneous adjustment is not supported, however, as the estimate of λ_{MInt} is significantly different from 1. The numerical value of λ_{PQ} also indicates that more than half of the adjustment in the price of gross output is completed within one year. It is interesting to observe that the mean adjustment time of the quantity appears to be shorter than of the price. But the null-hypothesis that the two estimates are equal is accepted at the 99% confidence level, employing either one of the two standard errors in the test. The combined rate of technological progress of domestic value added and imported intermediate goods is estimated to be relatively low (.59%) and not significantly different from zero.

Applying cost minimization to equation 6.2, the factor demand for the efficiency capital stock J can be derived as a function of domestic value added (Y) and the relative price between the two factor inputs, namely the wage rate (w) and the price of the efficiency capital stock (P_J):

$$(6.9)\quad J = \left[\frac{b}{1-b}\right]^{1-b} \left[\frac{w}{P_J}\right]^{1-b} Y.$$

If equation 6.3 is also subjected to the cost-minimization assumption, factor demands for K_d and K_f can be derived as a function of Y by substituting 6.9 in the first-order conditions of a minimum. This gives

$$(6.10)\quad K_d^* = \left[\frac{a}{1-a}\right]^{1-a} \left[\frac{P'_{MCap}\ E\ (1+t_{Cap})}{P_Q}\right]^{1-a} \left[\frac{b}{1-b}\right]^{1-b} \left[\frac{w}{P_J}\right]^{1-b} e^{(-z_K t)}\ Y$$

as the derived demand for the desired capital stock consisting of domestically manufactured capital goods and

$$(6.11)\quad K_f^* = \left[\frac{1-a}{a}\right]^{a} \left[\frac{P_Q}{P'_{MCap}\ E\ (1+t_{Cap})}\right]^{a} \left[\frac{b}{1-b}\right]^{1-b} \left[\frac{w}{P_J}\right]^{1-b} e^{(-z_K t)}\ Y$$

for the desired long run stock build from imported machinery. The user cost of the capital stocks is represented by the price deflator of new investment. The stock is evaluated at the replacement cost. For the domestic stock, this price is equal to the price of gross output (P_Q) since a one-sector economy is considered and the price of investment of the domestic output product therefore corresponds to the output price. The user cost of K_f is the border price of imported capital goods in domestic currency (P'_{MCap} E), adjusted for the tariff rate on imported machinery and equipment ($1+t_{Cap}$). The parameter measuring technological progress, z_K, is again a weighted average of the individual technological progress terms of the two capital stocks:

$$z_K = z_{Kd}a + z_{Kf}(1-a).$$

The (non-observable) price of the efficiency capital stock, P_J, is a weighted average of the shares of the two capital stocks according to

$$(6.12)\quad P_J = P_Q \left[\frac{K_d}{J}\right] + P'_{MCap}\ E\ (1+t_{Cap}) \left[\frac{K_f}{J}\right].$$

This yields the following price equation, if we substitute the derived factor demands of the production function (6.3) into equation 6.12:

$$(6.13)\quad P_J = P_Q^{\,a} \left[P'_{MCap}\ (1+t_{Cap})\ E\right]^{1-a} \left\{\left[\frac{1-a}{a}\right]^{a} + \left[\frac{a}{1-a}\right]^{(1-a)}\right\} e^{(-z_k t)}.$$

Rewriting equations 6.10 and 6.11 in log-linear form and incorporating a lagged adjustment process results in

$$(6.14)\quad \ln(K_d) = \lambda_{Kd}\,(1-a)\,\ln\left[\frac{a}{1-a}\right] + \lambda_{Kd}\,(1-a)\,\ln\left[\frac{P'_{MCap}\,E\,(1+t_{Cap})}{P_Q}\right]$$

$$+ \lambda_{Kd}\,(1-b)\,\ln\left[\frac{b}{1-b}\right] + \lambda_{Kd}\,(1-b)\,\ln\left[\frac{w}{P_J}\right] - \lambda_{Kd} z_K t$$

$$+ \lambda_{Kd}\,\ln(Y) + (1-\lambda_{Kd})\,\ln(K_{d,t-1}) + u_{14}$$

and

$$(6.15)\quad \ln(K_f) = \lambda_{Kf}\,a\,\ln\left[\frac{1-a}{a}\right] + \lambda_{Kf}\,a\,\ln\left[\frac{P_Q}{P'_{MCap}\,E\,(1+t_{Cap})}\right]$$

$$+ \lambda_{Kf}\,(1-b)\,\ln\left[\frac{b}{1-b}\right] + \lambda_{Kf}\,(1-b)\,\ln\left[\frac{w}{P_J}\right] - \lambda_{Kf} z_K t$$

$$+ \lambda_{Kf}\,\ln(Y) + (1-\lambda_{Kf})\,\ln(K_{f,t-1}) + u_{15}.$$

Since one main focus of our analysis is to examine how imports of capital goods behave under different growth and policy scenarios, it would suffice to estimate equation 6.15 separately. But the choice of K_d either as based on production statistics or on gross domestic investment was observed to influence the estimate of the elasticity of substitution between the two capital stocks in some cases (compare Section 5.3.5.).[9] Therefore, we estimate the two demand equations 6.14 and 6.15 jointly [10], including the definition of P_J as described in equation 6.13. Results are presented in Table 6.2. The two regressions produce very good fits and small standard errors. The share of the efficiency capital stock in valued added (*b*) is estimated to be around 60%, the labor share at 40%. The technological progress term z_k is found positive, as expected, but insignificantly

9 As total production of the economy is analyzed here, the broader of the two capital goods concepts, namely the one based on gross domestic investment expenditure, is employed in the estimations. See Appendix C, Section C.5.

10 See Coen and Hickman (1970).

Table 6.2:
Columbia: Joint NLLS estimation of the factor demands for the two capital stocks (K_d and K_f), 1972-1987

endogenous variable	$R^2_{adj.}$	Dh	SSE
$\ln (K_d)$	.994	-.436	.0163
$\ln (K_f)$	.984	1.602	.0057

parameter	estimate	stand. error	t-statistic
b	.5941	.2088	2.845
a	.8831	.0356	24.80
z_k	.0166	.0109	1.523
λ_{Kd}	.0784	.0383	2.047
λ_{Kf}	.3654	.0600	6.090

different from zero. According to the estimation, the long run share of imported capital stock in the efficiency stock is 12% (1-a), which underpins the dominant role indigenous investment goods play in total capital formation in Columbia. The mean adjustment of K_f is less than two years, while the capital stock made up of domestic capital goods requires a much longer time to adapt to its long run level.

Theoretically, the system of equations should have also incorporated the estimation of the GDP deflator (P_Y), since it is a weighted average of the factor shares of labor and the efficiency capital stock according to

$$(6.16)\quad P_Y = t_{PY} \left\{ P_J \left[\frac{J}{Y}\right] + w \left[\frac{L}{Y}\right] \right\},$$

with t_{PY} as an indexation effect due to the inclusion of the tariff on imported capital goods. After substituting the derived factor demands for J and L in equation 6.16, the following expression for the long run desired deflator of domestic value added can be obtained:

$$(6.17)\quad P_Y^* = t_{PY}\, P_J^{\,b}\, w^{1-b} \left\{ \left[\frac{b}{1-b}\right]^{1-b} + \left[\frac{1-b}{b}\right]^{b} \right\}.$$

Table 6.3:
Columbia: NLLS estimation of the GDP price deflator (P_Y) equation, 1972-1987

endogenous variable	$R^2_{adj.}$	serial	SSE
ln (P_Y)	.999	Dh=.296	.0110
parameters	**estimate**	**stand. error**	**t-statistic**
P_{Y0}	-.0851	.0270	-3.115
λ_{PY}	.6017	.1762	4.105

The log-linear form of this equation (including a price adjustment process similar to the one used in equation 6.8 above) becomes

$$(6.18)\ \ln(P_Y) = P_{Y0} + \lambda_{PY}\, b \ln(P_J) + \lambda_{PY}\,(1-b) \ln(w) + (1-\lambda_{PY}) \left[2 \ln(P_{Y,t-1}) - \ln(P_{Y,t-2})\right] + u_{18};$$

with

$$P_{Y0} = \lambda_{PY} \ln\left[t_{PY}\right] + \lambda_{PY} \ln\left\{\left[\frac{b}{1-b}\right]^{1-b} + \left[\frac{1-b}{b}\right]^{b}\right\}.$$

Incorporating 6.18 in the equation system of 6.13, 6.14, and 6.15 produces non-converging estimates. We employ a single-equation non-linear estimation instead, restricting the distribution parameter b to .5941, the estimate presented in Table 6.2 and defining P_J according to equation 6.13 (also using the paramter estimates presented in the table). The regression outcome is shown in Table 6.3. The adjustment time of the GDP deflator (λ_{PY}) is estimated to be much faster than the quantity adjustments (λ_{Kf} and λ_{Kd}).

Finally, the last category of imported goods which has to be determined consists of all other goods than capital and intermediate ones (M_{Oth}). It is postulated that the demand for these other imports is determined by their domestic price faced by the consumer relative to the domestic output price, and real income. In log-linear form this equation becomes

Table 6.4:
Columbia: OLS estimation of the demand for other imports (M_{Oth}), 1972-1987

endogenous variable	$R^2_{adj.}$	DW	SSE
$\ln(M_{Oth})$	.925	2.277	1.104

parameter	estimate	stand. error	t-statistic
M_{Oth0}	-6.690	1.984	-3.372
ψ_Y	1.558	.4181	3.726
ψ_P	-1.126	.0923	-12.20

$$(6.19)\quad \ln(M_{Oth}) = M_{Oth0} + \psi_P \ln\left[\frac{P'_{MOth}\, E\, (1+t_{Oth})}{P_Q}\right] + \psi_Y \ln(Y) + u_{19},$$

with P'_{MOth} as the dollar price index of the border price, t_{Oth} as the tariff on this import category, ψ_P as the price elasticity of demand, and ψ_Y as the respective income elasticity. Inclusion of a lagged adjustment mechanism results in an estimate of λ_{MOth}, the speed of adjustment of other imports, slightly above (but insignificantly different from) 1. The assumption of instantaneous adjustment is for that reason compatible with this estimation result. Regression results are shown in Table 6.4. Demand for the 'other' goods in Columbia is both income- and price-elastic.

Although the coefficient of determination is relatively high, the large standard error of the regression infers that prediction of this import category is quite difficult. It contains mostly consumer goods, which are often described as the import category showing the highest variation, because governments restrict them if intermediate and capital goods are exerting pressure on the balance of payments.[11] This role of consumer goods imports might explain that short run fluctuations are difficult to describe with a standard demand function as used here.

11 See Mirakhor and Montiel (1987, p.69) and Salas (1982, p.308).

For the simulation of import responses to external shocks, imports of capital goods (M_{Cap}) have to be deduced from the calculation of the desired capital stock. They are derived from the stock accounting identity:

$$(6.20)\quad M_{Cap} = K_f - K_{f,t-1} + \delta \left[\sum_{i=t-8}^{t-1} M_{Cap,i} \right],$$

where δ is the depreciation rate and assumed equal to .125 as in the reference estimations of Chapter 5. A linear depreciation method is again used with no diminution taking place in the period of investment. The term in brackets includes all past period investment that has not completely depreciated if such a function is postulated.

The value of total imports in *current dollars* (MV) is then represented by

$$(6.21)\quad MV = \frac{M_{Int}}{E_{80}} P'_{MInt} + \frac{M_{Cap}}{E_{80}} P'_{MCap} + \frac{M_{Oth}}{E_{80}} P'_{MOth},$$

where the constant 1980 pesos series of each import item is transformed into the constant 1980 dollar series by dividing by the nominal exchange rate in the base year (E_{80}).

6.2.1.2. Export demand and supply

The easiest (and standard) way of modelling exports is to assume that any one country faces a downward sloping demand curve for its exports and is itself able to supply unlimited amounts of exports, i.e. the supply elasticity is assumed to be infinite. But as already pointed out in the introduction to this chapter, the conditions that underlie such a supply function are unlikely to be fulfilled since it requires constant or increasing returns to scale. The postulation of an infinite elasticity of export supply would also connote that the small-country assumption of the import side is completely reversed; demand for a country's exports has no effect on the export price in such a case.[12]

The other alternative, namely to assume that the developing country faces given world prices for its exports, would imply (in the one-sector economy assumed here) that the domestic output price of the goods is determined by the price it can achieve for them on the export market.

---

12 See Siebert (1989, pp.90-91) for an analysis of export structures of the highly indebted countries and the effect of supply increases on the terms of trade.

But as Melo and Robinson (1989, p.50) put it, "it is also reasonable to assume that world prices in the tradable sectors do not dominate the domestic price system".

While staying within the concept of a one-sector economy, it is possible to differentiate between the domestic output price (P_Q) and the price the country can achieve for this good on world markets (P_X) by assuming product differentiation on the export side. Domestically produced goods sold on the domestic market are then imperfect substitutes for goods sold on the export market (Melo and Robinson 1989, p.50). With this concept in mind, an export supply function like

$$(6.22)\quad X_S = f_{XS}\left[\frac{P_X\ E}{P_Q},\ Q\right]$$

can be derived, where the supply of exports (X_S) is dependent on the real output capacity (Q) and the relative price between the one achieved on the world and the domestic market.[13] This functional form is deduced from an export transformation function according to which producers distribute their output on the export and domestic market. Rather than deriving explicit parameters for the elasticity of transformation (which would require a constant elasticity of transformation aggregation function), it suffices for the purpose at hand to assume a log-linear relationship, including a Koyck-geometric lag adjustment:

$$(6.23)\quad \ln(X_S) = \lambda_X X_{S0} + \lambda_X \varphi_P \ln\left[\frac{P_X E}{P_Q}\right] + \varphi_Q \lambda_X \ln(Q) + (1-\lambda_X)\ln(X_{S,t-1}) + u_{23}.$$

Both φ_P, the price-elasticity of export supply, and φ_Q, the output-elasticity of export supply, are expected to be positive.

13 Compare Melo and Robinson (1989, p.52). They specify the export supply equation as $X_S/D_S = g(P_Q, P_X)$, where D_S is the amount of output supplied on the domestic market. Similar to the derivation of factor demands, D_S is itself a function of X_S, Q, and relative prices, so that an equation of the form 6.22 can be derived. Compare e.g. Marquez (1984, appendix) for consecutive derivation of factor demands. Melo and Robinson (1989) introduce an export transformation function in a computable general equilibrium model for a small economy in order to distinguish between the domestic and export price for the one good produced in the economy. In these models, the price on the domestic market is determined endogenously by the interaction of supply and demand (derived from an import aggregation function). Since we do not look at the demand side explicitly (and therefore also consumption of the domestically produced goods), P_Q is determined by the mark-up hypothesis as outlined above.

Export demand is modelled as a function of the relative price between Columbia's exports and the GDP dollar price deflator of its major trading partners (P_W) and the latters' real income (Y_W).[14]

$$(6.24)\quad X_D = e^{X_{D0}} \left[\frac{P_X\ E}{P_W\ E}\right]^{\xi_{PW}} \left[Y_W\right]^{\xi_{YW}}.$$

The dollar unit value index of exports (P_X) is determined endogenously by the interaction of supply and demand. Solving equation 6.24 for the export price, including a price-adjustment process in line with the specification for P_Q and P_Y, and taking logs on both sides gives us

$$(6.25)\quad \ln(P_X) = -X_{D0}\ \lambda_{PX} \frac{1}{\xi_{PW}} + \frac{1}{\xi_{PW}} \lambda_{PX}\ln(X_D) - \lambda_{PX} \frac{\xi_{YW}}{\xi_{PW}} \ln(Y_W)$$

$$+ \lambda_{PX}\ln(P_W) + (1-\lambda_{PX})\left[2\ \ln(P_{X,t-1}) - \ln(P_{X,t-2})\right] + u_{25}.$$

Equating X_D and X_S

$$(6.26)\quad X = X_D = X_S$$

allows us to estimate 6.23 and 6.25 simultaneously with two-stage nonlinear least squares.[15] Estimation results are given in Table 6.5. In the supply equation, the output elasticity of exports is estimated at 1.717, an estimate not significantly different from 1 at the 5% level.[16] The price-elasticity of export supply has the expected positive sign and indicates a price-inelastic export

---

14 Columbia exports as much as 80% of total exports to the developed market economies, so that the OECD price deflator and OECD GDP data are used to proxy price and income developments among the major trading partners. See Appendix C.5.

15 For simultaneous estimation of export supply and demand functions compare Arize (1990), Bond (1987), and Goldstein and Khan (1978). Theoretically, the simultaneous estimation should be carried out employing a full-information method like three-stage least squares. But the application of this method to estimation of equations 6.23 and 6.25 does not improve parameter values.

16 If an export supply function is derived from the constant elasticity of transformation function, the expected parameter value of φ_Q is 1.

Table 6.5:
Columbia: Two-stage NLLS estimation of the export supply and demand functions (X_D and P_X), 1972-1987

endogenous variable	$R^2_{adj.}$	serial	SSE
ln (X)	.793	LM=.2247	.1242
ln (P_X)	.827	Dh=.0016	.3021

parameters	estimate	stand. error	t-statistic
X_{S0}	-17.31	10.59	-1.635
φ_P	.1873	.3119	.6005
λ_X	.5711	.2960	1.929
φ_Q	1.717	.5024	3.417
X_{D0}	-11.00	9.294	-1.184
ξ_{PW}	-.7345	.2938	-2.500
ξ_{YW}	1.181	.4404	2.682
λ_{PX}	.6452	.1270	5.080

supply function, but it is not found significantly different from zero.[17] This would indicate that export supply is largely determined by output capacity. This result may be partly due to the omission of export subsidies and taxes in the measurement of (P_XE), as this price should represent the price a domestic producer receives for exporting the domestically produced good. But time series

17 Donges and Riedel (1977, p.66) estimate an export supply function for Columbia and find the relative price variable to be insignificantly different from zero as is the case above. Bond (1987) finds her export supply function of specific commodities exported by the developing countries to yield relatively 'poor' results as the price-elasticity of export supply carries the correct sign and is significantly different from zero at the 10% level only in 6 out of 19 estimated export supply equations. Compare also Arize (1990), who examines export behavior in Asian developing countries and traces inelastic supply functions for India, Pakistan, the Philippines, and Thailand. The results indicate a price-elastic response of export supply for Indonesia and Malaysia only.

data for export taxes or a proxy thereof could not be obtained.[18] Further, institutional factors might lower the direct response of exports to changing prices since Columbia is one of the main world exporters of coffee and its export quantities have been strongly influenced by agreements of the International Coffee Agreement (ICA). The ICA was established in 1962, collapsed in 1972, but was revitalized to be in effect from October 1980 to February 1986. Since export quotas are fixed by the agreement, the response of coffee supply to changing prices is obviously limited if the cartel is operating and the quota agreements are obeyed.[19]

Parameters have the expected sign in the export demand function and they are significantly different from zero. According to these results, Columbia faces an income-elastic, but price-inelastic demand for its exports. While the explanatory power of the two equations is satisfactory, the standard error of regression in the price equation is relatively large. But estimation of export unit values for developing countries is generally accompanied by low coefficients of determination and high regression standard errors.[20]

Specifying valued exports (XV) in current dollars as

$$(6.27) \quad XV = \frac{X}{E_{80}} P_X$$

and the trade balance (TB) in current dollars

$$(6.28) \quad TB = XV - MV$$

closes the model.[21]

---

18 Additionally, the export unit value is also influenced by the real value of the dollar versus other currencies of trading partners. A significantly negative effect of a real dollar appreciation on commodity prices has been established in several studies (e.g. Cline 1984, Dornbusch 1985, Gilbert 1986, Goldsbrough and Zaidi 1986). This effect is not incorporated here either.

19 See Herrmann (1988) for a description and analysis of the International Coffee Agreement. Ronning (1991) analyzes West-German coffee imports and models the presence of binding export quotas by including a dummy variable in his estimation.

20 Compare Saunders and Dean (1986, p.170). See Goldsbrough and Zaidi (1986, pp.157-162) for a discussion of determinants of nominal and real commodity prices and a survey of empirical findings.

21 Including definitions and identities, the model is represented by equations 6.5, 6.8, 6.13, 6.14, 6.15, 6.18, 6.19, 6.20, 6.21, 6.23, 6.25, 6.26, 6.27, and 6.28.

Table 6.6:
Columbia: Evaluation of the model: ex post simulation, 1972-1987[1]

variable	Corr (1)	Theil's U (2)	ME/MAE (3)	RMSE/MAE (4)
K_f	.987	.0241	.0131	1.158
K_d	.998	.0091	.2430	1.217
M_{Int}	.944	.0576	.1948	1.158
M_{Oth}	.839	.1502	-.1442	1.234
P_Y	.999	.0410	-.6563	1.459
P_Q	.996	.0387	-.5894	1.340
X	.919	.0491	.1133	1.084
P_X	.737	.0809	.1245	1.215

1 *Corr* is the correlation between the actual and simulated value of the respective endogenous variable; *Theil's U* specifies the Theil inequality coefficient; *ME*, *MAE* and *RMSE* stand for the mean error, mean absolute error and root mean square error, respectively.

6.2.2. Ex post simulation

Beside the trend variable, the model contains ten exogenous variables: domestic and world GDP, three import prices and the three respective tariff rates, the domestic wage rate, and the exchange rate as a policy variable. Additionally, the dynamic adjustment process consists of six lagged price variables and four lagged quantity variables, hence ten lagged endogenous variables. We validate the model by dynamic ex post simulation over the period of estimation (1972-1987). Table 6.6 contains some statistics on how good the model tracks the actual development of the endogenous variables.[22] The first column gives the correlation coefficient between the actual and simulated values of the respective endogenous variable, which is generally high. Only the unit value index of exports and the 'other import'- category show a somewhat lower correlation, but this could already be expected from the regression results. The second column presents Theil's

22 See König (1990, pp.139-144), Koutsoyiannis (1977, pp.490-497), and Pindyck and Rubinfeld (1981, pp.360-374) for a description and explanation of statistics that evaluate how well a simulation reproduces actual values.

inequality coefficient.[23] As can be seen in the table, the inequality coefficient is relatively low for all variables, with highest values for the export price (P_X) and 'other imports' (M_{Oth}). Further, Table 6.6 contains two ratios in the last two columns. The first measures the mean error in relation to the mean absolute error. The mean error should be considerably smaller than the mean absolute error, since this would otherwise indicate that a systematic bias is present. The ratio should be small. This is the case for all variables except for the two domestic prices (P_Y and P_Q), which exhibit a relatively high value. The next column sets the root mean square error in relation to the mean absolute error. If the former is a multiple of the latter, a misspecification of the structural equation could be present.[24] No indication of such a misspecification is detected. Since the focus of the next subsection is to see how changes in the external environment influence the dollar trade balance, actual and simulated values of three variables are plotted in Figures 6.1 to 6.3. These are the dollar value of imports, the dollar value of exports, and the resulting trade balance.

While the dollar import values are reproduced quite nicely by the simulation, the export value more or less shows only the trend of export revenue developments and fluctuations around this trend are not picked up by the export specification. Nevertheless, the trade balance swings over the simulation period are reproduced with the account deteriorating till the beginning of the 1980s and improving thereafter.

6.2.3. Analyses of trade balance reactions

We now use the model to analyze how different environments influence the trade balance. The calculations are not designed to forecast endogenous variables (which would be a relatively hopeless endeavor considering the political situation in which Columbia finds itself today), but rather to see how the structural importance of imports (combined with the export specification) translates itself into trade balance reactions.

We conduct the computations in the following way: In the base scenario, import prices, tariffs, the domestic wage, and the exchange rate are kept at their 1987 level. World income (here represented by the respective OECD value) is increased by 2% over the following periods and so is the world (OECD) GDP deflator. Finally, domestic income is assumed to grow by 3% per year which allows us to calculate the resulting trade balances. Average five-period values of the three

23 Theil's inequality coefficient can range from 0 to 1. If it is 0, the simulated and actual values are identical; the higher the inequality coefficient, the worse the predictive performance of the model.

24 Compare König (1990, p.140).

Figure 6.1: Columbia: Simulated and actual values of the import value, 1973-1987

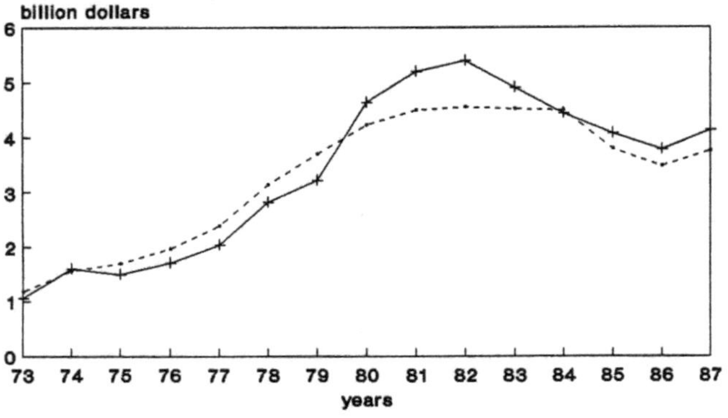

Figure 6.2: Columbia: Simulated and actual values of the export value, 1973-1987

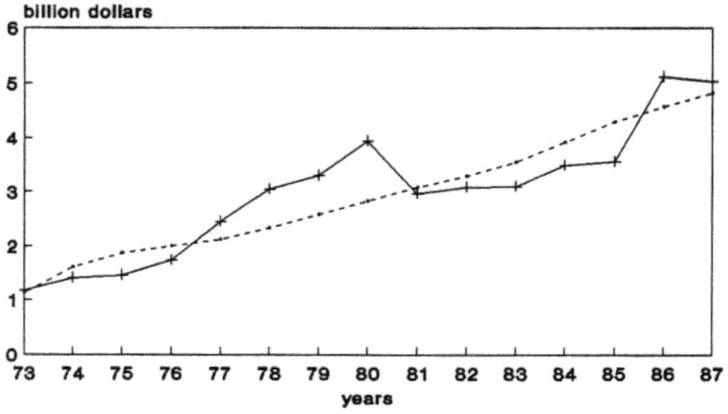

Figure 6.3: Columbia: Simulated and actual values of the trade balance, 1973-1987

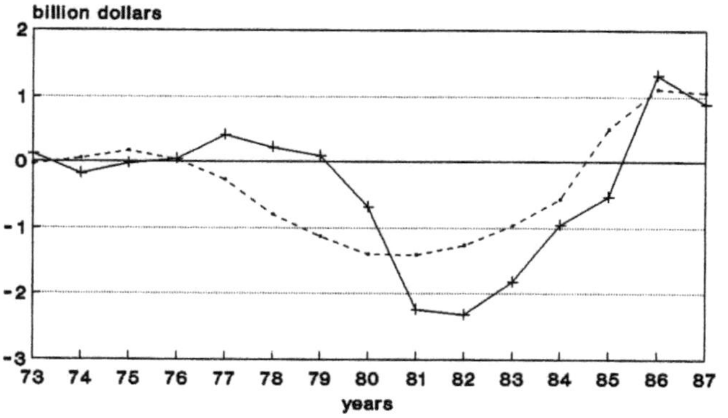

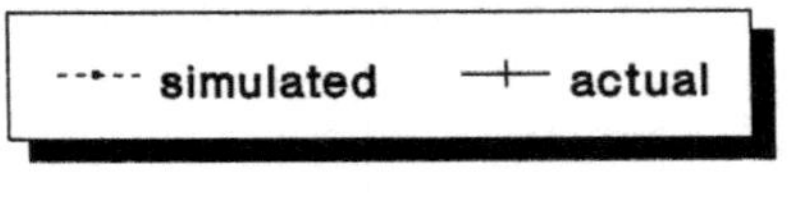

import quantities (M_{Cap}, M_{Int}, and M_{Oth}), export volume (X_Q), and the three endogenous prices (P_Q, P_Y, and P_X) are then computed, excluding the first period (in which changes are introduced in the other scenarios). In Table 6.7 the first column contains indices of these variables (set to 1.0 in the base scenario) as well as the resulting values of exports, imports, and the trade balance (in current dollars). The five-period average trade balance that would result under these assumptions comes close to one billion dollars, which is about the value actually achieved in 1987.

In the first variation, we introduce a nominal devaluation of 15% with the exchange rate kept at this level thereafter; the other exogenous variables remain as in the base scenario.[25] The rise in the nominal exchange rate causes an increase in the local currency price of imported capital and intermediate goods. Imported and domestic factors are substitutes in production and this stimulates a switch in factor input combinations in favor of the capital stock consisting of domestic investment goods and labor; the quantity of imported means of production is reduced in response to the change in relative prices. The second column in Table 6.7 shows that the volume of both imported capital and intermediate goods decreases by about 10% compared to the base result. Although the entire system of production is described by Cobb-Douglas functions, the percentage decrease in import demand for these two categories is less than the intial percentage change in the exchange rate. This might be due to the specification of a dynamic adjustment process, which causes a slow-down in the reaction of the trade balance. But the contribution of this phenomenon is of only minor importance as the period in which the exchange rate is modified is excluded from the five-period averages in Table 6.7 and the mean adjustment time (in which half of the desired change takes place) is estimated at .5 years for imported intermediate goods (Table 6.1) and less than 2 years for the capital stock made up of imported machinery (Table 6.2). More important is the fact that the rise of the import prices (in pesos) causes both the gross output deflator (equation 6.8) as well as the domestic GDP deflator (equation 6.18) to increase since imported capital and intermediate goods are used as factors of production. As reported in Table 6.7, P_Y - the domestic GDP deflator - increases on average by 3.5% over the base value and P_Q even by 4.5%. The relative price change of imported compared to domestic input factors is dampened; the nominal devaluation of 15% translates into a real devaluation of 10.5% (using the output deflator as the reference price).

The quantity of other imports (M_{Oth}) decreases more than for capital and intermediate goods since instantaneous adjustment to the alternated relative price occurs and the demand for

---

25 The nominal wage rate is also kept as in the base case since no feed back mechanism between a change in the price of gross output and the nominal wage is included. This would require the incorporation of the price index of consumer goods and expectations thereof.

Table 6.7:
Columbia: Simulation results for domestic GDP growth of 3%[1]

variable	base (1)	devaluation (2)	tot decline (3)	OECD growth (4)
P_Y	1.000	1.035	1.046	1.000
P_Q	1.000	1.045	1.059	1.000
P_X	1.000	1.003	1.032	1.062
X_Q	1.000	.998	.976	1.010
M_{Int}	1.000	.901	.873	1.000
M_{Cap}	1.000	.907	.881	1.000
M_{Oth}	1.000	.848	.806	1.000
XV	5480	5483	5530	5884
MV	4484	4023	4672	4484
T_B	996	1460	858	1399

1 Indices (solution of the 'base' simulation is set to 1); for the value of exports (XV), the value of imports (MV) and the trade balance (TB) current US dollars. Entries represent five-period average values (the year in which the variation is introduced is not taken into account).

this product category is price-elastic (equation 6.19 and Table 6.4). The dollar value of total imports decreases substantially by slightly more than 10%.

On the export side, the nominal devaluation has two opposing effects on export supply. First, the domestic price of exports increases relative to the domestic output price. This leads to an increase in the quantity of exports and puts downward pressure on the export unit value index (in dollars). These quantity and price effects are relatively small since the export supply function is price inelastic. Second, as imported intermediate inputs decrease in this scenario relative to the base one, gross output of the economy falls marginally, which has a negative effect on export supply and consequently a positive one on the export unit value index. Demand for Columbian exports does not change compared to the base scenario. The combined effect of the above mentioned opposing influences on the supply function reduces the export quantity marginally, while the export unit value index is raised slightly. Export receipts remain more or less at the base case value, so that the strong improvement in the trade balance (more than 45%) as a consequence of the devaluation is attributable to the decline in import volumes of all three product classes.

The second variation examines the effect of a terms of trade decline. A once and for all 20% increase in the import prices is analyzed, which is assumed to relate to all three import categories alike.[26] Results of this modification are reported in the third column of Table 6.7. Import volumes of intermediate, capital, and 'other' goods decline beyond the level resulting from a 15% devaluation since the price increase of imported goods (in domestic currency) is higher. Consequently, prices of domestic value added and the output price are driven up more (4.6% and 5.9%). As the nominal exchange rate is kept constant, the increase in the output price induces an appreciation of the real exchange rate, causing the export quantity to decline. The latter effect is complemented by the relative decline in output capacity as imports of intermediate goods fall compared to the base case. Since demand for exports is inelastic, the increase in the export unit value index overcompensates the drop in export volume. The dollar value of exports rises marginally. The dollar value of imports (MV) increases by only 4.2% in response to a 20% import price shock. It can be attributed to the considerably lower import demand for capital and intermediate goods since these are substitutes for domestically produced investment goods and labor. The deterioration of the trade balance is consequently relatively small.

The third variation introduces an exogenous rise in export demand via higher world income (OECD) growth (4% instead of 2% in the other scenarios). The outward shift of export demand causes both the dollar price and the quantity of exports to increase; again, the quantity effect is minor as the export supply function is price-inelastic. The strong rise in export prices above the base period evolution (+6.2%) increases export earnings strongly.[27] As neither the import prices nor the nominal exchange rate deviate from their respective paths in the base case, import quantities, import values, and domestic prices correspond to the base scenario. The trade balance improves primarily due to the strong rise in the dollar unit value index of exports. However, the trade balance amelioration falls short of the one achieved with a 15% devaluation of the nominal exchange rate.

In order to visualize the effect of the diverse variations of the exogenous variables for different domestic growth rates, we compute the trade balance for GDP growth rates ranging from 0 to 10%. Each of the four scenarios (base plus three modifications) results in trade balance computations as a function of the postulated growth rate of domestic GDP. The relationship between these two variables is reproduced in Figure 6.4. Entries again represent five-period aver-

---

26 Both a devaluation and a change of the terms of trade cause the domestic prices of imports to increase. Imports are substituted for by the domestic product in consumption and production. But the dollar price of imports increases in the terms of trade scenario only, so that the effect of a negative shock on the trade balance can be studied.

27 Compare Golstein and Khan (1985, p.1089) for a discussion of the role of price elasticities in export supply and demand functions.

ages with the first one (the period of change) excluded from the calculation. The shift in the 'growth - trade balance curve' in response to changing external conditions is of interest here rather than the position of it, which links specific growth rates and trade balance computations.

The observations made when examining the findings on how the trade balance changes in case of a 3% domestic growth are reproduced if the domestic income path is varied. A terms of trade deterioration (20% rise in import prices across the board) has a negative effect on the trade balance, but it is relatively small in magnitude when compared to the effect of a 15% devaluation of the nominal exchange rate, which moves the 'growth - trade balance curve' outwards considerably. While the positive effect of a higher OECD income growth balances the effect of the devaluation at lower growth rates, it falls short of the latter if income growth is higher. This is mainly due to the fact that the export value decreases slightly with higher growth rates since the outward shift of supply (due to higher domestic supply capacity) is accompanied by overproportionally lower export unit values due to the inelastic nature of world demand. For high domestic growth rates, the positive influence of the higher world demand in the third modification increases the export value in absolute terms less than the devaluation reduces the import bill.

Figure 6.4: Columbia: Simulating trade balance reactions

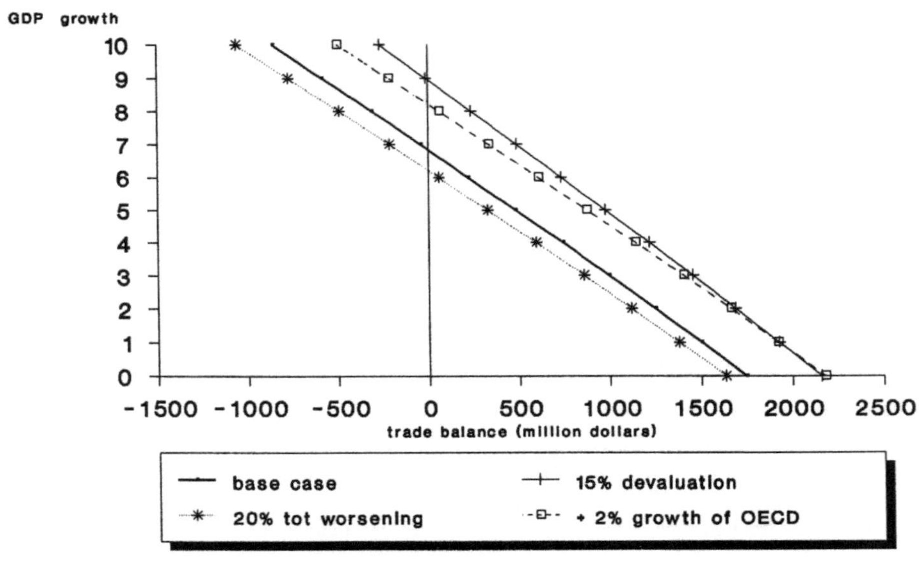

The overall picture that emerges is that Columbia, which is characterized by significant substitution possibilities for the main import goods (capital and intermediate ones), is able to influence its (dollar) trade balance significantly by an exchange rate devaluation. Similarly, the flexibility of domestic production enables the economy to absorb a terms of trade deterioration without experiencing a marked negative impact on the trade balance. These same issues are now examined and quantified for Ecuador.

6.3. The Ecuadorian model

6.3.1. Econometric specification

As in the Columbian model, the Ecuadorian equations are specified and estimated in the national currency; all quantities (including imports and exports) are specified in 1980 sucres. The trade balance is again expressed in current dollars.

6.3.1.1. Factor demands, price equations, and the demand for other imports

The estimations of the elasticity of substitution between the domestic capital stock and that consisting of imported machinery and equipment in the previous chapter produced low and insignificant values. This gives support to the assumption that the two capital stocks are employed in fixed proportions in Ecuador.[28] The elasticity of substitution between domestic value added and imported intermediate inputs, on the other hand, was found both significantly different from zero and 1, so that the more general CES-formulation is retained at the hyper level. Finally, the elasticity of substitution between labor (L) and the efficiency capital stock (J) is again assumed to be equal to unity in the long run, which allows us to approximate this function by a Cobb-Douglas type. The three functions are:

$$(6.29)\quad Q = \left[c \left[e^{z_Y t} Y\right]^{-\tau} + (1-c) \left[e^{z_M t} M_{Int}\right]^{-\tau} \right]^{-1/\tau} \qquad \text{with } \sigma_m = 1/(1+\tau)$$

$$(6.30)\quad Y = J^{b} L^{1-b}$$

28 The interpretation of the estimated substitution elasticities follows Lucas (1989, p.145), who employs the fixed proportions approach if the estimate of the elasticity of substitution of the CES-function is insignificantly different from zero.

$$(6.31)\ J = \min \left[\frac{K_d\ e^{z_{Kd}t}}{a}, \frac{K_f\ e^{z_{Kf}t}}{(1-a)} \right].$$

σ_m is the elasticity of substitution between imported intermediate inputs and domestic value added, z_i represents disembodied factor-specific technological progess. The two stocks K_d and K_f form the efficiency capital stock J according to a Leontief production function. Following Farzin (1988, p.10), we can specify this function in a way so that a and (1-a) represent the shares of the domestic capital stock and the one made up of imports in the efficiency stock.

Assuming cost-minimizing firms, the factor demand for imported intermediate inputs from the CES hyper-function 6.29 becomes

$$(6.32)\ M_{Int}^{*} = \left[\frac{1-c}{c}\right]^{\sigma_m} \left[\frac{P_Y}{P'_{MInt}\ E\ (1+t_{Int})}\right]^{\sigma_m} Y\ e^{(z_Y-z_M)t\tau\sigma_m}.$$

E is the exchange rate index (1980=1) of sucres per dollar and t_{Int} represents the tariff rate on imported intermediate inputs. Including a partial adjustment process and taking logs on both sides yields the following estimating equation:

$$(6.33)\ \ln(M_{Int}) = \lambda_{MInt}\ \sigma_m\ \ln\left[\frac{1-c}{c}\right] + \lambda_{MInt}\ \sigma_m\ \ln\left[\frac{P_Y}{P'_{MInt}\ E\ (1+t_{Int})}\right]$$

$$+ \lambda_{MInt}(1-\sigma_m)(z_Y-z_M)t + \lambda_{MInt}\ln(Y)$$

$$+ (1-\lambda_{MInt})\ \ln(M_{Int,t-1}) + u_{33},$$

with λ_{MInt} as the speed of adjustment. After substituting the derived factor demands into the cost function 6.6, the output price P_Q becomes

$$(6.34)\ P_Q^{*} = (1+r)\ t_{PQ} \left\{ P_Y^{(1-\sigma_m)}\ c^{\sigma_m} e^{z_Y t(\sigma_m-1)} \right.$$

$$\left. + \left[P'_{MInt}E(1+t_{Int})\right]^{(1-\sigma_m)} (1-c)^{\sigma_m} e^{z_M t(\sigma_m-1)} \right\}^{\frac{1}{1-\sigma_m}}$$

with r as the mark-up, which is modelled analogue to the Columbian model; t_{PQ} is again a constant term due to the indexation effect of inclusion of the tariff rate on imported intermediate inputs. Equation 6.34 incorporates the rates of technological progress as specified in equation 6.29 in the cost specification of CES production functions derived by Uzawa (1962) and Sato (1967).[29]

Specifying a price adjustment mechanism in which the change in P_Q is linked to the desired and past alterations in prices, gives

$$(6.35)\quad \ln(P_Q) = P_{Q0} + \frac{1}{1-\sigma_m}\lambda_{PQ}\ln\Big\{P_Y^{(1-\sigma_m)}\, c^{\sigma_m} e^{z_Y t(\sigma_m-1)}$$

$$+ \Big[P'_{MInt}E(1+t_{Int})\Big]^{(1-\sigma_m)}(1-c)^{\sigma_m}\, e^{z_M t(\sigma_m-1)}\Big\}$$

$$+ (1-\lambda_{PQ})\Big[2\ln(P_{Q,t-1}) - \ln(P_{Q,t-2})\Big] + u_{35},$$

where

$$P_{Q0} = \lambda_{PQ}\ln(1+r) + \lambda_{PQ}\ln(t_{PQ}),$$

as the estimating equation.

We estimate equations 6.33 and 6.35 jointly with non-linear least squares; results are reported in Table 6.8. The distribution parameter *c* is calculated to be 98.4%, much higher than the corresponding result for Columbia. As outlined in the data appendix, only the Ecuadorian imports from the developed market economies are taken into account here. But since Ecuador is a net exporter of oil, these trade flows cover a very high share of its total intermediate imports.[30]

29 See Appendix B.

30 Ecuador reported imported goods on a detailed SITC level only till 1984 at the time these estimations were conducted. Trade flows had to be reversed, using the developed market economies as reporters. Since Ecuador is an oil-exporting country, close to total imports of intermediate goods should be covered. Ecuador imports about 80% of its total imports from the developed market economies (UNCTAD 1990, p.124).

Table 6.8:
Ecuador: Joint NLLS estimation of the factor demand for intermediate imports (M_{Int}) and the output price equation (P_Q), 1973-1987

endogenous variable	$R^2_{adj.}$	serial	SSE
ln (M_{Int})	.885	LM=3.080	.0700
ln (P_Q)	.987	Dh=-.1476	.0160

parameter	estimate	stand. error	t-statistic
c	.9837	.0231	42.582
σ_m	.5751	.1742	3.301
P_{Q0}	-.0549	.0593	-.926
z_Y	.0052	.0052	1.000
z_M	.0746	.0510	1.463
λ_{PQ}	.8609	.1234	6.977
λ_{MInt}	.7315	.2756	2.654

The long run elasticity of substitution is, as in Chapter 5, both significantly different from zero as well as from 1. As we already observed in the Columbian estimations and in the various regression results reported in the last chapter, the mean adjustment time which intermediate imports require to adjust to their optimal level is very short, here estimated at less than half a year, whereas the price-adjustment of P_Q to changes in its components is almost instantaneous. Disembodied technological progress takes place at a fast pace in regard to imported intermediate goods and to a much lower degree in regard to domestic value added. Both estimates of the technological progress terms are insignificantly different from zero.

Turning to the demand for the two capital goods, firms facing a fixed proportions production function as specified in equation 6.31 attempt to minimize production costs (subject to an efficiency capital stock constraint) by employing the two capital stocks so that no idle capacity of either stock exists:[31]

---

31 Compare Intriligator (1978, p.273).

$$(6.36a)\quad K_d^{*} = \frac{\bar{J}\, a}{e^{(z_{Kd} t)}}$$

$$(6.36b)\quad K_f^{*} = \frac{\bar{J}\,(1-a)}{e^{(z_{Kf} t)}}$$

If 6.36a and 6.36b are now substituted in the demand for the efficiency capital stock, J, which corresponds to the model specification for Columbia (equation 6.9), the optimal demand for each capital stock becomes in logarithmic form (including the usual partial adjustment formulation):

$$(6.37)\quad \ln(K_d) = \lambda_{Kd}\,(1-b)\,\ln\left[\frac{b}{1-b}\right] + \lambda_{Kd}\,(1-b)\,\ln\left[\frac{w}{P_J}\right] - \lambda_{Kd} z_{Kd} t$$

$$+ \lambda_{Kd}\,\ln(a) + \lambda_{Kd}\,\ln(Y) + (1-\lambda_{Kd})\,\ln(K_{d,t-1}) + u_{37},$$

$$(6.38)\quad \ln(K_f) = \lambda_{Kf}\,(1-b)\,\ln\left[\frac{b}{1-b}\right] + \lambda_{Kf}\,(1-b)\,\ln\left[\frac{w}{P_J}\right] - \lambda_{Kf} z_{Kf} t$$

$$+ \lambda_{Kf}\,\ln(1-a) + \lambda_{Kf}\,\ln(Y) + (1-\lambda_{Kf})\,\ln(K_{f,t-1}) + u_{38}.$$

The lagged response characterizes the limited flexibility of the capital stocks and permits idle capacity to be present. The specification in equations 6.37 and 6.38 allows us to estimate the parameter *a*, which determines the composition of the efficiency capital stock J. Since the elasticity of substitution between the two investment goods is zero, the relative price between the two does not influence the factor demands. The price of the efficiency capital stock P_J, which is not observable, becomes

$$(6.39)\quad P_J = P_Q\, a\, e^{(-z_{Kd} t)} + \left[P'_{MCap}(1+t_{MCap})E\right]\,(1-a)\, e^{(-z_{Kf} t)}.$$

Further, the GDP price deflator equation contains some of the parameters included in 6.37-6.39 as well. Since the specification of the value added function corresponds to the form for Columbia, the price deflator equation matches equation 6.18:

Table 6.9:
Ecuador: Joint NLLS estimation of the factor demands for the two capital stocks (K_d and K_f) and the GDP price deflator equation (P_Y), 1973-1987

endogenous variable	$R^2_{adj.}$	serial	SSE
$\ln(K_d)$	.973	Dh=1.014	.0496
$\ln(K_f)$	.948	Dh=1.575	.0301
$\ln(P_Y)$	.998	DW=2.415	.0087

parameter	estimate	stand. error	t-statistic
a	.6402	.0906	7.066
b	.7198	.0671	10.73
z_{Kd}	-.0220	.0085	-2.588
z_{Kf}	.0836	.0257	3.253
λ_{Kd}	.0972	.0391	2.486
λ_{Kf}	.2090	.0604	3.460
P_{Y0}	.0103	.0495	.2081

$$(6.40)\quad \ln(P_Y) = P_{Y0} + \lambda_{PY}\, b\, \ln(P_J) + \lambda_{PY}\,(1-b)\,\ln(w) + (1-\lambda_{PY})\left[2\ln(P_{Y,t-1}) - \ln(P_{Y,t-2})\right] + u_{40};$$

with

$$P_{Y0} = \lambda_{PY}\ln\left[t_{PY}\right] + \lambda_{PY}\ln\left\{\left[\frac{b}{1-b}\right]^{1-b} + \left[\frac{1-b}{b}\right]^{b}\right\}.$$

The system of equations 6.37 to 6.40 contains three endogenous variables (P_Y, K_f, K_d) and one definition (P_J). The regression produces an estimate of λ_{PY}, the adjustment parameter of the GDP price deflator, of slightly (but insignificantly) above 1, so that we assume complete adjustment of the price variable within one year (λ_{PY} is restricted to 1). The statistics of the resulting estimation are summarized in Table 6.9. The explanatory power of each equation is very high and the statistics measuring the presence of autocorrelation are satisfactory.

We estimate the distribution parameter in the fixed-proportions function, *a*, at .64, which emphasizes the significant role imported capital goods play in capital formation in Ecuador. The long run share of the value of the efficiency capital stock in nominal value added (*b*) is 72%, also higher than for Columbia, where it was slightly above 60%. The important role of imported capital goods in the aggregate capital stock and the dominance of the latter in the formation of value added combine to underpin the major role attached to imports of machinery and equipment in Ecuador. Corresponding to the Columbian case, adjustment of the imported capital stock is faster than the one for domestic capital goods. Finally, a high rate of disembodied technological progress is attached to the imported capital stock and the parameter is found highly significant. On the other hand, the rate of technological progress associated with the domestic capital stock K_d is determined to be significantly negative.[32]

The third and last import category concerns other imports, which are supposed to be a function of the relative price between them and the domestic goods as well as domestic income. But estimation results are unsatisfactory. Autocorrelation occurs, which might be due to the presence of a structural break. The Durbin-Watson statistic improves significantly when an intercept-shifting dummy variable is included, which takes on the value of 1 from 1982 onwards (the debt crisis years) and is 0 otherwise. In addition to the tariff included on other imports, the dummy variable is designed to account for the impact of import restrictions which were imposed with the tightening of the international financial markets. Again, this product category consists largely of consumption goods, which have been found to react most strongly if a foreign exchange shortage occurs.[33] The equation to estimate becomes

$$(6.41)\quad \ln(M_{Oth}) = M_{Oth0} + \psi_P \ln\left[\frac{P'_{MOth}E(1+t_{Oth})}{P_Q}\right] + \psi_Y \ln(Y) + d_M D_M + u_{41},$$

with $$D_M = \begin{cases} 0 \text{ if year} \leq 1981 \\ 1 \text{ if year} \geq 1982. \end{cases}$$

32 See Chenery et. al. (1986, Chapters 8 and 10) for a discussion of technological change, total factor productivity growth, and their determinants. Michalopoulos (1975, p.353) states that inefficient import substitution policies in domestic capital goods production can cause declining productivity levels of domestically originating investment (which can show up as a negative rate of technological progress).

33 Compare Hentschel (1989), Mirakhor and Montiel (1987), and Salas (1982).

Table 6.10:
Ecuador: OLS estimation of the demand for other imports (M_{Oth}), 1973-1987

endogenous variable	$R^2_{adj.}$	serial	SSE
ln (M_{Oth})	.788	DW=2.457	.1769

parameter	estimate	stand. error	t-statistic
M_{Oth0}	-5.322	6.183	-.861
ψ_Y	1.143	.316	3.617
ψ_P	-.765	.255	-3.000
d_M	-.606	.098	-6.184

Regression results are presented in Table 6.10. We find Ecuadorian demand for this product category to be income-elastic (as for Columbia), but price-inelastic. The parameter measuring whether a structural break occured in 1982 (d_M) is highly significant, indicating that actions were taken to reduce the demand for 'other imports' after the outbreak of the crisis.

The two closing identities complete the determination of the import side (derivation of capital goods imports from the stock accounting identity and specification of valued imports in dollars) and they are equivalent to equations 6.20 and 6.21.

6.3.1.2. Export supply and demand

Export supply is related to the relative price between the price received on the world and the domestic market and output capacity. Including a partial adjustment function, the equation to estimate becomes

$$(6.42)\quad \ln(X_S) = \lambda_X X_{S0} + \lambda_X \varphi_P \ln\left[\frac{P_X E}{P_Q}\right] + \varphi_Q \lambda_X \ln(Q) + (1-\lambda_X)\ln(X_{S,t-1}) + u_{42},$$

which is identical to equation 6.23. But the specification of the demand function varies slightly from the one used for Columbia. Ecuador experienced a sharp drop of 25% in its export prices in 1986. This price decrease was primarily due to the sharp drop in oil prices as Saudi-Arabia increased its production of crude oil when abandoning its function as the swing producer in OPEC.[34] This drop is taken into account by including an intercept-shifting dummy variable (D_{PX}) in the export demand specification. Further, demand for Ecuadorian exports is postulated to be a function of world (OECD) income and the relative price between the Ecuadorian exports and the world price level (as in equation 6.24). Solving the demand function for the export unit value index, including the dummy variable, introducing a partial adjustment process in accord with the other price functions, and taking logs on both sides produces:

$$(6.43)\quad \ln(P_X) = - X_{D0}\,\lambda_{PX}\,\frac{1}{\xi_{PW}} + \lambda_{PX}\,d_{PX}\,D_{PX} + \frac{1}{\xi_{PW}}\,\lambda_{PX}\,\ln(X_D)$$

$$- \lambda_{PX}\,\frac{\xi_{YW}}{\xi_{PW}}\,\ln(Y_W) + \lambda_{PX}\,\ln(P_W)$$

$$+ (1-\lambda_{PX})\left[2\,\ln(P_{X,t-1}) - \ln(P_{X,t-2})\right] + u_{43}$$

with $$D_{PX} = \begin{cases} 0 \text{ if year} \leq 1985 \\ 1 \text{ if year} \geq 1986 \end{cases}$$

and d_{PX} as the parameter included to account for the sharp decline in export prices in 1986.

Equating X_D and X_S

$$(6.44)\quad X = X_D = X_S$$

renders simultaneous estimation of 6.42 and 6.43 possible. Regressions produce an estimate of φ_P, the price-elasticity of supply, which carries the expected positive sign, but it is (as in the Columbian case) insignificantly different from zero. Additionally, neither λ_X (the adjustment parameter of supply), nor φ_Q (the output elasticity of supply) are significant. The relative price term in 6.42 is therefore dropped from the equation; supply is specified here to depend solely on output capacity.[35]

---

[34] See Rauscher (1989) for a review and discussion of oil-price fluctuations and OPEC behavior.

[35] The determination of export supply by gross output only is not unusual in theoretical and empirical modelling of developing countries exports. Compare Farzin (1988), McKinnon (1964), and Takagi (1981).

Table 6.11:
Ecuador: Two-stage NLLS estimation of the export supply and demand functions (X_D and P_X), 1973-1987

endogenous variable	$R^2_{adj.}$	Serial	SSE
$\ln(P_X)$	.836	DW=2.073	.1539
$\ln(X)$	.467	Dh=.079	.1142

parameters	estimate	stand. error	t-statistic
X_{S0}	1.235	9.276	.1331
λ_X	.4675	.2023	2.311
φ_Q	.8698	.4771	1.823
X_{D0}	-1.952	6.620	-.2948
ξ_{PW}	-.6194	.1502	-4.123
ξ_{YW}	.7634	.2550	2.994
d_{PX}	-.4210	.1866	-2.256

Further, the adjustment of export prices turns out to take place within one year, as the estimate of λ_{PX} is slightly above one;[36] complete adjustment of the export unit value within one year is hence assumed. Including these restrictions ($\varphi_P=0$, $\lambda_{PX}=1$), two-stage nonlinear least squares is applied to equations 6.42 and 6.43. Estimates of the remaining parameters are shown in Table 6.11. While the coefficient of determination is satisfactory for the price-equation, the explanatory power of export supply is limited. Further, the output elasticity of export supply ($\varphi_Q=.87$) is only marginally significant at the 10% level. A reformulation of the export supply function including both domestic policies targeted at influencing the price of exports and institutional factors which dominated a large part of Ecuadorian exports (OPEC, ICA) might improve these results. But they are beyond the scope of this exercise, which is primarily designed to examine the behavior of the three different import types and their link to domestic production.

36 Bond (1987) specifies that export demand adjusts within a period of a year and export supply requires more time. The above finding that demand adjusts faster than supply for both Columbia and Ecuador is in accordance with her study.

Table 6.12:
Ecuador: Evaluation of the model (ex post simulation), 1973-1987[1]

variable	Corr (1)	Theil's U (2)	ME/MAE (3)	RMSE/MAE (4)
K_f	.933	.0484	-.2758	1.234
K_d	.947	.0505	-.1809	1.177
M_{Int}	.940	.0266	.3786	1.318
M_{Oth}	.910	.0553	-.0358	1.109
P_Y	.995	.0347	-.0823	1.311
P_Q	.990	.0481	-.0336	1.185
X	.697	.0517	-.0158	1.232
P_X	.692	.1026	-.1381	1.230

1 *Corr* is the correlation between the actual and simulated value of the respective endogenous variable; *Theil's U* specifies the Theil inequality coefficient; *ME*, *MAE*, and *RMSE* stand for the mean error, mean absolute error, and root mean square error, respectively.

The parameter estimates of the export demand equation carry the expected signs. Demand for Ecuadorian exports appears to be both income- and price-inelastic; the slope shifting dummy, which is incorporated to account for the large drop in oil prices in 1986, has the expected negative sign and its parameter estimate is significantly different from zero.

The specification of valued exports (in dollars) and the trade balance (in dollars) carries over from the Columbian closure (equations 6.27 and 6.28).[37]

6.3.2. Ex post simulation

The goodness of fit of the model is examined by tracking the endogenous variables in a dynamic simulation over the period of estimation (1973-1987). Statistics of this exercise are presented in Table 6.12. The correlation between the simulated and the actual values of the endogenous variable is high in all cases except for the export quantity and the export price, where it is

37 The complete Ecuadorian model (including identities and definitions) consists of equations 6.20, 6.21, 6.27, 6.28, 6.33, 6.35, 6.37, 6.38, 6.39, 6.40, 6.41, 6.42, 6.43, and 6.44.

only about 70%. Since the export quantity (fixed in the supply equation alone) enters into determination of the export price, any simulation errors of the former carry over in the export price function.

Theil's inequality coefficient is small for all variables except for the export unit value index (P_X). As desired, the mean error is only a fraction of the mean absolute error (column three), while the root mean square error is not a multiple of the mean absolute error (column four).

Actual and simulated values are plotted for the value of imports and exports (in current dollars) and the dollar trade balance in Figures 6.5 to 6.7. Both values and turning points of actual total imports of Ecuador are reproduced by the dynamic simulation. This statement holds for valued exports to a somewhat lesser extent (especially from 1979 to 1982), but the large fluctuation of export receipts from 1982 to 1987 is tracked very closely. The actual trade balance showed minor fluctuations around a balanced account (1974-1982), significant and rising surpluses (1982-1985), and a subsequent decline (1985-1987). These are reproduced by the simulation; less well in regard to the small movements at the beginning of the period and close to perfect in the second half.

6.3.3. Analyses of trade balance reactions

We can now use the Ecuadorian model to examine how the three different scenarios (devaluation, terms of trade decline, OECD growth) affect the trade balance of this country, which is characterized by a strong link between domestic economic performance and imported factors of production. Although the Ecuadorian supply-side specification contains a fixed-proportions production function linking domestically produced and imported capital goods, substitution possibilities for both imported factors of production are present in the model: labor and imported intermediate inputs can be employed to replace the capital stock from imports (K_f) and both types of capital stocks as well as labor substitute for imported intermediate inputs. The supply-side specification incorporates much more flexibility than rigid fixed proportions models.

In the base run, OECD income and price paths correspond to their respective developments in the Columbian base scenario, being increased by 2% per period each, domestic income (GDP) is postulated to grow by 3%. The domestic wage rate, import prices, and the three tariff rates are kept at their 1987 level. Entries in Table 6.13 all represent five-period average values, not taking the first simulation period into account. This base scenario results in a hypothetical trade balance surplus of 700 million dollars as shown in Table 6.13. Imports, exports, and the trade balance are expressed in current million dollars, while indices are used for all other variables. They are normalized to 1 in the base run.

Figure 6.5: Ecuador: Simulated and actual values of the import value, 1974-1987

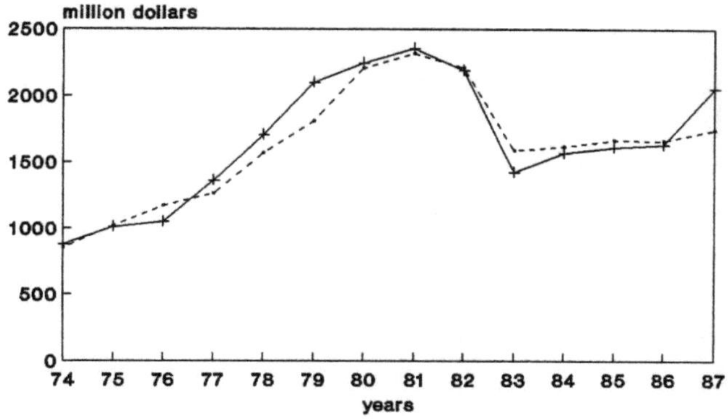

Figure 6.6: Ecuador: Simulated and actual values of the export value, 1974-1987

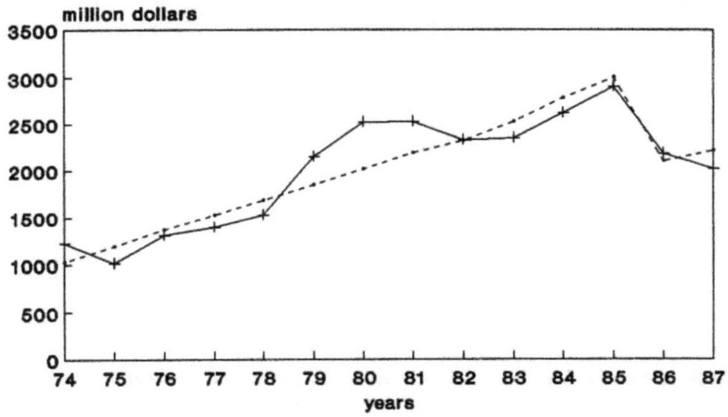

Figure 6.7: Ecuador: Simulated and actual values of the trade balance, 1974-1987

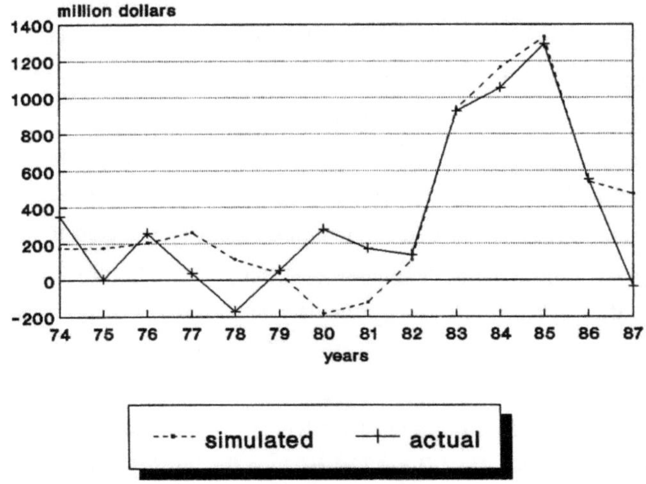

Table 6.13:
Ecuador: Simulation results for domestic GDP growth of 3%[1]

variable	base (1)	devaluation (2)	tot decline (3)	OECD growth (4)
P_Y	1.000	1.044	1.058	1.000
P_Q	1.000	1.051	1.068	1.000
P_X	1.000	1.009	1.010	1.062
X_Q	1.000	.995	.994	1.000
M_{Int}	1.000	.946	.931	1.000
M_{Cap}	1.000	.986	.982	1.000
M_{Oth}	1.000	.924	.908	1.000
XV	2600	2609	2610	2762
MV	1940	1836	2167	1940
T_B	660	773	443	822

1 Indices (solution of the 'base' simulation is set to 1); for the value of exports (XV), the value of imports (MV), and the trade balance (TB) current US dollars. Entries represent five-period average values (the year in which the variation is introduced is not taken into account).

The first variation examines the effect of a 15% nominal devaluation while the other exogenous variables are kept at their base case values. The higher relative price between imports and their domestic substitutes reduces demand for all three import categories, but to a different degree. The smallest reduction occurs (as expected) for imported capital goods since substitution possibilities are limited and adjustment to desired levels takes a considerable period of time (mean lag of four years). Imports of machinery and equipment are estimated to fall short of their base case volume by 1.4% only. A somewhat higher degree of substitutability (and shorter adjustment period) is associated with imports of intermediate inputs, whose import quantity is calculated to decline by 5.4% in response to the devaluation. The category of imports affected most are 'other goods', which are determined in the consumption decision and found to respond to price changes within a period of a year. Imports of this category are curtailed by 7.6% in response to the 15% nominal devaluation. The effect is not higher since the import price increase (in domestic currency) pushes up the GDP-price deflator (P_Y) by an average of 4.4% and the gross output price by 5.1%, so that the 15% nominal devaluation comes down to a real devaluation of close to 10%. The higher share of and lower substitutability for imported capital goods leads to a higher rise of domestic prices in Ecuador as compared to Columbia.

A minor reduction of gross output occurs due to lower imports of intermediate goods. As export supply is solely a function of gross output, it is affected only marginally. The resulting increase in the export unit value index is consequently marginal as well. The total import value decreases by slightly more than 5% which translates (with basically constant export quantity and export unit value) into a rather small improvement in the trade balance.

The second variation inspects the outcome of a once and for all increase in import prices by 20% across the board. The results are reported in the third column of Table 6.13. Quantities of all import categories are suppressed below the level experienced when examining the effect of the devaluation, since the import price rise causes a more pronounced change in the relative prices between imports and domestic means of production. Most striking is again the small variation of capital goods imports, which we calculate to fall short of the base-run value by only 1.8%. The supply-inflationary effect drives up the price of value added by 5.8% and of gross output by 6.8%. Export quantity and price show little reaction again. The import price rise increases the import bill substantially, since the major import categories (intermediate and capital goods) respond only little. The trade balance deteriorates strongly by 30% as compared to the base scenario.

Finally, the third variation (column 4 in Table 6.13) concerns the impact of higher world growth rates on the Ecuadorian trade balance. The import side and the domestic prices are not altered in relation to the base case. As export supply is independent of prices, export quantities do not respond to the higher world demand and lead to an increase (due to the output capacity effect) in line with the base projection. The entire impact of higher world growth is transmitted to an increasing export unit value index which rises 6.2% above its respective value in the reference calculation. The trade balance is activated by about 150 million dollars, still more than if a 15% devalution is considered.

Using these scenarios but varying the growth of domestic income produces a relationship between the trade balance and the domestic growth rate as depicted graphically in Figure 6.8. By far the largest alteration of the growth-trade balance link occurs in the case of a terms of trade deterioration, specified as a 20% rise in import prices. Correspondingly, the positive effect of a devaluation on the trade balance is relatively small. Although the empirical examination of the export side could not establish a link between the export quantity and world demand, the upward

Figure 6.8: Ecuador: Simulating trade balance reactions

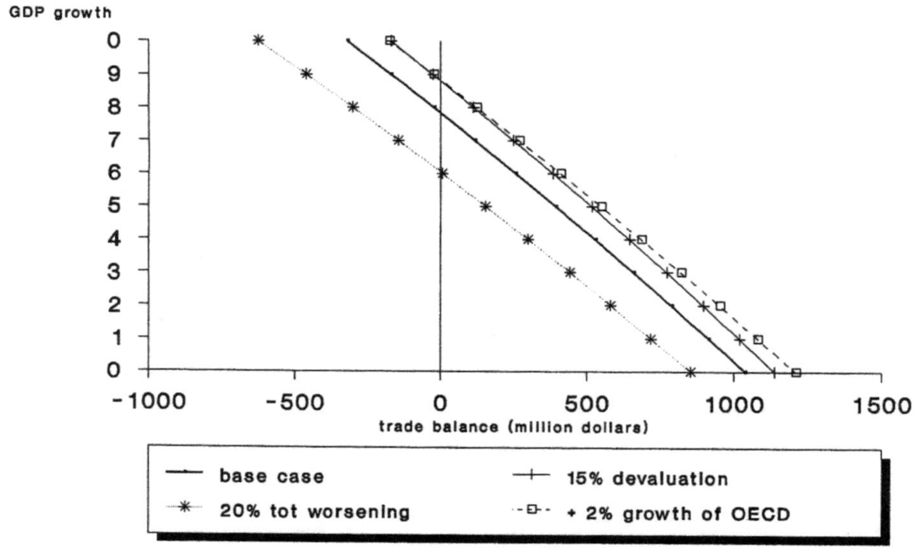

pressure on the export unit value index is sufficient to raise export revenues more than if a 15% nominal devaluation occurs.

It emerges that Ecuador is vulnerable to external shocks. The applicability of devaluation policies is limited on account of the high importance of imported capital and intermediate goods and the price-inelastic export supply. Similarly, the trade balance worsens considerably in response to a sudden increase in import prices. Deflationary policies constitute the main adjustment mode to achieve a trade account impovement.

6.4. Concluding remarks about the simulations

In this chapter, we introduced a simple trade balance model that allowed us to quantify the impact of the importance of imported intermediate and capital goods in developing economies. Two countries were chosen that had been found to be characterized by diverging supply structures, Columbia and Ecuador. Factor demands of the imported means of production were derived for cost minimizing firms. The specification of cost functions allowed us to endogenize some relevant domestic prices. An attempt was also made to describe the export market by modelling both supply and demand. But a significant link between export supply and a relative price of the exports to the domestic price could not be established for either country. It is possible that this phenomenon can be attributed to the omission of domestic policies, such as subsidies or export taxes, which influence the domestic price exporters receive. Additionally, the major share of the quantity of exports of both countries was determined by commodity agreements (OPEC, ICA) for a considerable part of the period that we employed for estimation. Quantities of the major exports were predetermined by these agreements. The price responsivenes of exports was consequently limited to non-traditional goods. An extension of the framework, which takes into account both of these shortcomings relating to the export supply function, would strengthen the predicative power of the model.

Tracking of the models showed that the partial equilibrium framework traced the flow of merchandise trade and trade balance positions well. The simulations examined how the structural importance of imports influences the reaction of the trade balance (specified in current dollars) to a nominal devaluation, a terms of trade deterioration, and an increase in world demand.

In the Ecuadorian case, the limited substitution possibilities for both capital and intermediate goods render devaluation policies far less effective than in a country that is characterized by a weaker structural importance of imported factors of production, such as Columbia. Due to low substitution elasticities in production, the demand for imported factors of production responds little to changes in relative prices. Similary, unfavorable exogenous import price movements put strong pressure on the trade balance in Ecuador since a restructuring of factor inputs is possible only to a limited extent. Import volumes decline only marginally while the import bill rises strongly. For Columbia, on the other hand, import quantities decrease substantially in response to the price change. The negative effect of the terms of trade deterioration is mitigated.

Although the nominal wage was kept constant in the simulations, the import price increase and devaluation resulted in a significant acceleration of inflation solely due to the role of imported means of production. This supply-inflationary effect erodes nominal devaluation policies. It was found to be more pronounced for Ecuador.

Faced with the need to achieve a balance of payments improvement, countries which are characterized by strong structural linkages between imported goods and domestic productive capacity have to pursue growth-reducing policies in order to achieve a trade balance surplus unless a devaluation stimulates export performance. Chapter 5 concluded that policies which restrict imports of machinery in order to support the domestic capital goods sector did not raise substitution abilities for this product category. A dependence of growth on imports therefore calls for an active policy to stimulate traditional and non-traditional exports. This enables the financing of necessary imported inputs from export proceeds rather than from resource transfers from abroad. It results in more independence of the domestic economy as foreign exchange to pay for necessary imported inputs can be financed even if international credits cannot be attained.

7. Conclusion

7.1. Summary of findings

This study explored the role of imports as factors of production in twelve developing economies. All countries studied here experienced real import declines in the 1980s, when balance of payments difficulties were encountered. As consumer goods had accounted for only a minor position in the total import bill, the potential cushion provided by them was too small to avoid a decline of imported intermediate and capital goods.

A lot of studies assign the slackening growth of developing countries in the 1980s to this compression of imported means of production. It is postulated that the developing economies were dependent on these imported capital and intermediate goods as no close domestic substitutes were available. A supply side shock stemming from the shortfall of foreign-made machinery and intermediate inputs is made responsible for the observed output disturbances. The aim of this study was to examine the assumption about zero or low substitutability between imported capital and intermediate goods on the one hand, and domestic investment products and labor on the other.

A three-level nested production function was specified. Under the assumption that firms minimize production costs, we estimated short and long run elasticities of substitution between input pairs. The basic result obtained in these calculations is that significant substitution elasticities for both imported machinery and intermediate inputs exist in most developing countries studied here, which invalidates the fixed-proportions approach linking imports and output. Factor-input combinations react to relative price changes even in the short run, but the calculated elasticities of substitution characterize the stock of imported machinery as complementary to the stock consisting of domestically produced investment goods when a period of one year only is considered.

A lack of ability to replace imported investment goods even in the long run was only traced for Ecuador and the Philippines, both countries with comparably small activity in indigenous manufacture of machinery and equipment. No or only very weak evidence of significant substitution elasticities for intermediate imports was deduced for Brazil and the Philippines - economies that rely heavily on imports of primary inputs. While for most countries instantaneous adjustment of imported intermediate goods could be observed, the adjustment time of the capital stocks to their desired levels is significant. The average time required at which half of such a change is completed takes about two to three years.

The basic model used in the estimations of the elasticities of substitution employed a protection proxy to represent the impact of trade policies on the domestic price of imported goods. An extension introduced a government reaction function, linking the combined effect of tariff and non-tariff barriers to the availability of foreign exchange. The results we presented in this study suggest that most governments allocate scarce foreign exchange to intermediate goods and levy tariffs and quantitative restrictions primarily on imported machinery. The discriminatory distribution of foreign exchange accounts for the observed overproportional compression of capital goods imports in the 1980s.

The computed long run elasticities of substitution varied among the group of developing countries, which limits the applicability of world models describing the 'South' as a homogeneous group. A partial equilibrium model of trade and growth served to visualize the effects of structural differences relating to imported factors of production on trade balance reactions. If a country is dependent on imports as inputs into production (Ecuador was used as an example) and at the same time has limited export capabilities (either due to a limited diversification, protection of partner countries, or an anti-export bias), the main adjustment to a sudden foreign exchange shortage works mainly via growth reducing policies. Devaluations of the exchange rate help little to achieve a balance of payments amelioration as substitution possibilities for imports are limited and a large share of the initial nominal devaluation is eroded by a subsequent domestic price push stemming from the supply side. Similarly, a sudden increase in import prices is absorbed much less favorably in import-dependent economies than in those which do possess domestic substitutes for imported production means. Structural differences stemming from the use of imports should be taken into account when designing adjustment and policy packages for indebted economies. While 'adjustment-cum-growth' is possible for some countries, there is a trade-off between the two targets in economies which rely to a higher degree on imported factors of production.

The dispersion of estimated substitution elasticities was less pronounced for imported intermediate and more pronounced for imported capital goods. Studies on developed countries often characterize oil and other intermediate inputs as complements in the production process. The findings of this study are in accord with this assessment in regard to the substitutability for imported intermediate goods in developing countries. Non-traded factors of production, which largely cause trade in primary inputs, offer one possible explanation for the relatively low variance of calculated substitution elasticities for this product group.

Contrary to this finding relating to imported intermediates, the calculated elasticities of substitution between the capital stock made up of domestically produced machinery and that of imported investment goods showed a considerable dissimilarity. We tested whether this divergence can be explained by indicators relating to domestic capital goods production in the developing economies. A strong positive link between the strength of the capital goods sector (measured as the contribution of manufacturing value added of the capital goods industries to total GDP) and calculated elasticities of substitution was established. Since machinery producing activities are highly technology- and research-intensive, the spread of technological know-how stemming from activity in this sector creates indigenous skills which ease production of previously imported capital goods.

But the more important result is that an increased self-reliance on domestically manufactured machinery and equipment does *not* go hand-in-hand with higher 'independence', i.e. higher elasticities of substitution. Secondary import-substitution in capital goods, which restricts imports of machinery in order to create a market for domestic industries, has not achieved its primary aim of reducing the dependence on imported capital goods. Although the share of imported machinery in capital stock formation was reduced, they nevertheless remained vital for the functioning of the domestic economy. Substitutability could not be raised as the emerging industries themselves required imported investment and intermediate goods.[1]

These results are in line with the analysis of Jacobsson (1988) who observes that while a lot of developing countries have implemented import substitution policies in order to increase domestic manufacture of capital goods, the developed market economies have increasingly relied on intra-industry trade in capital goods in order to reap the benefits stemming from economies of specialization in this sector. A self-reliant capital goods sector is thus costly in two respects: First, developing countries become *technology-dependent* instead of independent since they are detached from international technological progress. This in turn hampers the development of indigenous technological capabilities and has a negative impact on productivity growth.[2] Second, as the economically efficient limit to import substitution is relatively low in the capital goods industries, prices have to be distorted drastically in order to stimulate domestic manufacture of previously imported goods, which causes an inefficient allocation of resources.

---

1 Compare Cardoso and Fishlow (1989a) and Little et al. (1970).

2 Dahlman et al. (1987, p.759) note that "some countries have overemphasized the drive for technological self-sufficiency and paid a high price in poor productivity".

7.2. Import vulnerability reconsidered

Michaely (1984, pp.7-8) defines the concept of economic vulnerability in the following way:

> *"Vulnerability is a function of two components. One is the extent of the damage that would occur should the disruption of the phenomenon on which the agent is dependent take place; the other is the likelihood of such an event, that is, of the disruption actually occurring".*

This vulnerability concept implies that countries which pursue a development strategy that attempts to strengthen domestic capital goods industries by restricting imports of investment goods become *more* vulnerable to a foreign exchange shortfall. "The self-sufficient economy, if it is underdeveloped and has not generated the capacity to transform itself, is in fact critically dependent on the outside world when it cannot produce the goods it needs", (Roemer 1981, p.430). An increased self-reliance in the supply of investment goods is not accompanied by higher substitution elasticities for imported machinery, but the anti-export bias which results from non-neutralized protectionist policies reduces the latitude for adjusting to a balance of payments disequilibrium. Since relatively low short run substitution elasticities mean that economies rely on their external sector to transform domestic savings into investment, the vulnerability of the domestic economy rises as the likelihood of an import shortfall is increased. Instead of being able to pursue the easier road of expenditure switching, much more costly growth reducing policies have to be enacted in order to compress imports, as exports cannot be expanded.

As credit availability for developing countries today is largely dependent on debt indicators such as the debt-export ratio, an anti-export bias can create a vicious circle of declining imports and output. Exports cannot be raised easily since they have been discouraged; thus new credit is not available. The burden of adjustment lies on the import side, but compressing imports means compressing output as short run substitution elasticities for imported means of production are low. Lower output levels, in turn, can erode the already weak export base.

A development strategy which does not bias incentives against export production therefore promises to decrease import vulnerability. This argument does not imply that countries pursuing an outward-oriented development strategy necessarily command higher elasticities of substitution between imported and domestic factors of production - this question requires further research. But they are less vulnerable to an import shortfall since export proceeds can be relied on to finance necessary imports.

Table A.1: Elasticities of substitution between the two capital stocks and selected indicators of captital goods production

Country	σ_R [1]	$\sigma_{.10}$ [2]	σ_{GDI} [3]	σ_{FX} [4]	Range of activities[5]	Average share of imported capital goods in apparent consumption[6] (1972-1987)	Average share of GDP originating in capital goods industries (1972-1987)
Argentina	1.835	1.374	1.296	1.378	20	.0644	.0523
Brazil	-[7]	-[7]	1.133	.5663[8]	92	.1216	.0769
Chile	1.287	.8797	.1857	.6644	26	.4390	.0249
Columbia	1.268	.9342	.6960	1.748	66	.4070	.0223
Ecuador	.1233	.1445	-[7]	.0058	n.a.	.6820	.0217
Indonesia	.5361	.4450	.7755	.5539	37	.6690	.0152
Malaysia	1.690	1.064	1.056	1.303	9	.7040	.0485
Mexico	.6597	.5024	.3910	.9157	31	.3760	.0401
Peru	.7896	.6883	.3903	1.138	39	.4210	.0286
Philippines	.3421	.3545	.5093	.5783	38	.6890	.0271
Uruguay	1.510	.8576	.8209	1.182	n.a.	.2570	.0228
Venezuela	.1858	.1540	.1957	.1888	17	.4830	.0216

Source: Tables 5.6, 5.7, 5.9, 5.11, and UNIDO (1988b, pp.126-127). For other data sources see Appendix C.

1 Calculated elasticities of substitution, reference estimations (Table 5.6).

2 Calculated elasticities of substitution, modification 1 estimations (depreciation rate 10%, Table 5.7).

3 Calculated elasticities of substitution, modification 3 estimations (gross domestic investment, Table 5.9).

4 Calculated elasticities of substitution, modification 4 estimations (foreign exchange availability, Table 5.11).

5 The number of activities of the capital goods sector refers to the year 1980 and is reported in UNIDO (1988b, pp.126-127).

6 Average share of imports of capital goods in apparent consumption of capital goods (production minus exports plus imports), 1972-1987 (for Mexico 1972-1986). The definition of capital goods production and imports follows Table C.1 in Appendix C.

7 The long run elasticity of substitution is not computed as the short run elasticity carries the wrong sign.

8 The capital stock derived from the gross domestic investment data is used for Brazil as it turned out that the capital stock from capital goods production resulted in a wrong sign for the substitution elasticities in the reference and modification 1 (depreciation rate 10%) estimations. This value is therefore excluded in the Spearman rank correlation analysis reported in Tables 5.12 and 5.16.

Appendix B. The cost function and conditional factor demands for a CES production function

The purpose of this appendix is to derive the conditional factor demands for a CES production function. We assume that firms minimize production costs subject to a given output level. Further, the conditional factor demands are used to deduce the cost function corresponding to the CES production function.

B.1. Derived factor demands

Consider the Ecuadorian hyper function which is introduced and estimated in Chapter 6:

$$(B.1)\quad Q = \left[c \left[e^{z_Y t} Y\right]^{-\tau} + (1-c) \left[e^{z_M t} M_{Int}\right]^{-\tau} \right]^{-1/\tau}$$

Q	gross output
Y	domestic value added
M_{Int}	imported intermediate goods
z_i	disembodied technological progress in Y and M_{Int}
c	distribution parameter
τ	substitution parameter
σ_m	elasticity of substitution between domestic value added and imported intermediate products with $\sigma_m = 1/(1+\tau)$.

The Lagrangian of equation B.1 is

$$(B.2)\quad \mathbf{L}_Q = P_Y Y + P_{MInt} M_{Int} + \Omega_Q \left\{ \bar{Q}_t - \left[c\left[e^{z_Y t} Y\right]^{-\tau} + (1-c)\left[e^{z_M t} M_{Int}\right]^{-\tau} \right]^{-1/\tau} \right\},$$

with P_Y and P_{MInt} as the prices of domestic value added and imported intermediate products, measured in the same currency. The tariff rate on imports of intermediate products is omitted for notational simplification. For cost minimizing firms the first order conditions are described by

$$(B.3)\quad \frac{\partial \mathbf{L}_Q}{\partial M_{Int}} = P_{MInt} - \Omega_Q \frac{\partial Q}{\partial MInt} = 0$$

$$(B.4)\quad \frac{\partial \mathbf{L}_Q}{\partial Y} = P_Y - \Omega_Q \frac{\partial Q}{\partial Y} = 0$$

$$(B.5)\qquad \frac{\partial L_Q}{\partial \Omega_Q} = \bar{Q} - \left[c\left[e^{z_Y t}Y\right]^{-\tau} + (1-c)\left[e^{z_M t}M_{Int}\right]^{-\tau}\right]^{-1/\tau} = 0$$

Combining equations B.3 and B.4 gives

$$(B.6)\qquad \frac{Y}{M_{Int}} = \left[\frac{c}{1-c}\right]^{\sigma_m}\left[\frac{P_{MInt}}{P_Y}\right]^{\sigma_m} e^{(z_M-z_Y)t\sigma_m\tau}.$$

Substituting equation B.6 in the condition B.5 gives

$$(B.7)\qquad M_{Int} = (1-c)^{1/\tau}\,\bar{Q}\,e^{-z_M t}\left\{\left[\frac{c}{1-c}\right]^{\sigma_m}\left[\frac{P_Y}{P_{MInt}}\right]^{1-\sigma_m} e^{(z_M-z_Y)t\sigma_m\tau} + 1\right\}^{1/\tau}$$

and

$$(B.8)\qquad Y = c^{1/\tau}\,\bar{Q}\,e^{-z_Y t}\left\{\left[\frac{c}{1-c}\right]^{-\sigma_m}\left[\frac{P_Y}{P_{MInt}}\right]^{\sigma_m-1} e^{(z_Y-z_M)t\sigma_m\tau} + 1\right\}^{1/\tau}.$$

which are the conditional factor demands for imported intermediate inputs and domestic value added.

B.2. The cost function

Output prices (P_Q) are assumed to be formed with a constant mark-up over production costs:

$$(B.9)\qquad P_Q = (1+r)\left\{P_Y\left[\frac{Y}{Q}\right] + P_{MInt}\left[\frac{M_{Int}}{Q}\right]\right\}.$$

Substituting the derived factor demands B.7 and B.8 in equation B.9 and rearranging terms yields:

$$\text{(B.10)}\quad P_Q^{1-\sigma_m} = \left\{ \left[\frac{c}{1-c}\right]^{-\sigma_m} \left[\frac{P_Y}{P_{MInt}}\right]^{\sigma_m-1} e^{(z_Y-z_M)t\sigma_m\tau} + 1 \right\}$$

$$\left\{ (1+r)^{(1-\sigma_m)} P_Y^{(1-\sigma_m)} c^{\sigma_m} e^{-z_Y t\sigma_m\tau} \right\}.$$

Raising both sides to the $1/(1-\sigma_m)$ and collecting terms gives the following cost function:

$$\text{(B.11)}\quad P_Q = (1+r)\left\{ P_Y^{(1-\sigma_m)} c^{\sigma_m} e^{z_Y t(\sigma_m-1)} + P_{MInt}^{(1-\sigma_m)} (1-c)^{\sigma_m} e^{z_M t(\sigma_m-1)} \right\}^{\frac{1}{1-\sigma_m}}.$$

Equation B.11 expresses the domestic output price as a function of the prices of the two input factors. It is the dual cost function to the CES production function in equation B.1.

Appendix C. Data sources and computational notes

C.1. Industrial production and price statistics

All industry-specific data (output and manufacturing value added in nominal and constant 1980 prices, national currencies) originate from the United Nations Industrial Development Organization Data Base (Global Studies). If necessary, they are converted into United States dollar series at constant 1980 prices with the 1980 nominal exchange rate of the respective currency versus the US dollar (IMF 1990c). The classification of industries producing mainly capital goods and industrial supplies is a compilation of information from UNIDO (1974, p.19; 1979, pp.172-173; 1983, p.141). It is presented in Table C.1.

The implicit price index of domestically produced capital goods (used in the estimations of Chapters 5 and 6) is derived from the nominal and constant national currency series. If necessary, it is converted into a US dollar index with the nominal exchange rate (IMF 1990c) and normalized to one in the base year (1980).

The share of manufacturing value added in GDP stems from the World Bank (1989d).[1]

Table C.1:
Classification of industries and matching of trade and industrial statistics

Industries by end-use	International statistics industrial classification	Standard international trade classification (Revision 1)
Industries producing mainly capital goods	381, 382, 383, 384, 385	69, 71, 72, 73, 81, 86 less 863
Industries producing mainly supplies and intermediates	321, 331, 341, 351, 352, 353, 354, 355, 356, 362, 369,371, 372	2312, 243, 25, 2626, 2627, 2628, 2634, 266, 332, 5, 62, 63, 64, 65, 661, 662, 663, 664, 665, 67, 68

1 The share of manufacturing value added in GDP is used to calculate indicators, which are pre-sented in Tables 2.2 and A.1, Figure 2.1, and employed in the rank correlation analysis (Table 5.16) in Chapter 5.

C.2. Trade data

Trade data according to the Standard International Trade Classification (SITC) stem from the United Nations Statistical Office (*Comtrade*). *Comtrade* contains a matrix of trade values by country, time period, and commodity.

Two different classifications into end-use categories are applied. The first is used when industry and trade data are compared or both used in an estimation. The matching between industrial and trade statistics is based on UNIDO (1979, pp.172-173) and it is included in Table C.1.[2]

The second classification takes into account all traded commodities and is not restricted to manufactures. It follows United Nations (1971) and is presented in Table C.2.[3] Two SITC headings are not assigned to a subgroup by United Nations (1971), namely motor spirit (SITC 3321) and motor cars (SITC 7321). Motor spirit can be both an intermediate product for production as well as a consumption good. Similarly, motor cars can acquire the role of a consumption or capital good. The values of these two SITC headings are distributed into end-use categories by the relative weight of the latter, excluding the two values: 'Motor spirit' is split up between intermediate and consumer goods, while the 'motor cars' value is distributed between capital and consumer goods.

In order to attain imports in constant 1980 prices, each SITC entry of Tables C.1 and C.2 is deflated by the corresponding 1-digit SITC import deflator. A maximum of nine different import deflators are applied to the trade data of each country and each year. The country-specific deflators originate from the UNCTAD Database on Price Deflators (Money, Finance, and Development Division). It is assumed that import prices of products within a 1-digit SITC-group move in line with each other. Deflated import values are aggregated in the same way as the nominal values. Implicit import price deflators for the end-use categories can then be computed from nominal and real import values.

Some of the developing countries contained in this study do not report trade data to *Comtrade* continuously till 1987. All trade flows used in the estimations concerning merchandise trade in capital goods are therefore reversed and the developed market economies (DMECs) are used

---

2 It is used in Tables 2.4, 2.6, A.1, and estimations of the elasticity of substitution between the capital stocks (Chapter 5), except for the modification 3 estimations.

3 This trade data definition is used in Table 2.1, the modification 3 estimations in Chapter 5, the estimations including intermediate imports in Chapter 5, and all estimations including trade quantities and trade prices in Chapter 6.

Table C.2:
Trade data classification into consumer, intermediate, and capital goods

Products by end-use	Standard international trade classification (Revision 1)
Capital Goods	692, 6982, 7111, 7112, 7113, 7116, 7117, 7118, 712, 7141, 7142, 7143, 71491, 715, 7171, 7172, 718, 7191, 7192, 7193, 71951, 71952, 71953, 7196, 7198, 722, 7249, 726, 7295, 7296, 7297, 72991, 72992, 72993, 72994, 72995, 72999, 7311, 7312, 7313, 7314, 7315, 7316, 7322, 7323, 7324, 7325, 7333, 7341, 73491, 7353, 7359, 8121, 82102, 86132, 86133, 86134, 86139, 86169, 8617, 8618, 86191, 86192, 86193, 86194, 86195, 86196, 86197, 86198, 86423, 86424, 8945, 89511, 89713, 89957.
Non-energy intermediate goods	0011, 0012, 0013, 0014, 0019, 041, 0421, 043, 044, 045, 046, 047, 0482, 04881, 04883, 052, 0536, 05462, 0548, 0554, 0611, 0615, 0619, 06202, 0711, 072, 08, 09906, 11211, 121, 21, 22, 23, 24, 25, 26, 27, 28, 291, 2921, 2922, 2923, 2924, 2925, 2926, 2929, 3326, 3329, 41, 4212, 4213, 4214, 4217, 422, 43, 51, 52, 531, 532, 5331, 5332, 53331, 53332, 53334, 53335, 5411, 5413, 5414, 5415, 5416, 54199, 5511, 55121, 55122, 55123, 5542, 56, 5711, 5712, 5713, 58, 59, 611, 6121, 6123, 613, 621, 6291, 6294, 62999, 631, 6321, 6322, 6324, 6328, 633, 641, 6421, 6423, 6429, 65111, 65112, 65113, 65115, 65121, 65122, 65123, 65124, 6513, 65141, 65151, 65153, 65161, 65162, 65164, 65171, 65172, 65174, 6518, 6519, 652, 653, 654, 655, 6561, 6578, 661, 662, 663, 664, 6651, 6658, 667, 67, 68, 691, 693, 694, 695, 69602, 69605, 69607, 6981, 6983, 6984, 69853, 6986, 6988, 69891, 69892, 69893, 69894, 69895, 69896, 69897, 7114, 7115, 71492, 71954, 7197, 7199, 723, 72912, 7293, 7294, 72996, 72998, 7317, 7326, 7327, 7328, 73292, 73312, 73492, 7358, 8122, 8123, 81241, 84154, 8611, 86199, 8623, 86241, 86243, 86244, 86245, 86413, 86414, 86421, 86425, 86426, 86429, 89112, 89143, 8919, 89291, 89292, 89433, 89512, 89591, 89593, 89594, 89595, 89921, 89925, 89927, 89943, 89952, 89953, 89955, 89991, 89992, 89994.
Energy-intermediates	32, 331, 3322, 3323, 3324, 3325, 34, 35.
Consumer goods	01, 02, 03, 0422, 0481, 0483, 0484, 04882, 051, 0532, 0533, 0535, 0539, 0541, 0542, 0544, 0545, 05461, 0551, 0555, 0612, 0616, 06201, 0713, 073, 074, 075, 091, 09901, 09902, 09903, 09904, 09905, 09907, 09909, 111, 11212, 11213, 1122, 1123, 1124, 122, 2927, 4215, 4216, 53333, 5417, 54191, 55124, 553, 5541, 5543, 5714, 6122, 6129, 6293, 62998, 6327, 6422, 65114, 65125, 65142, 65152, 65163, 65165, 65173, 65175, 6562, 6566, 6569, 6574, 6575, 6576, 6577, 6652, 666, 69601, 69603, 69604, 69606, 697, 69851, 65852, 69898, 7173, 7194, 7241, 7242, 725, 72911, 7292, 73291, 73311, 7334, 81242, 81243, 82101, 82103, 82109, 83, 8411, 8412, 8413, 8414, 84151, 84152, 84153, 84159, 8416, 842, 85, 8612, 86131, 8614, 8615, 86161, 86242, 863, 86411, 86412, 86422, 89111, 8912, 89141, 89142, 8918, 8921, 8922, 8923, 8924, 89294, 89299, 893, 8941, 8942, 89431, 89432, 8944, 8952, 89592, 896, 89711, 89712, 89714, 8972, 8991, 89922, 89923, 89924, 89926, 8993, 89941, 89942, 89951, 89954, 89956, 8996, 89993, 89995, 89996, 89997.
Other Goods[1]	0015, 73519, 89293, 89998, 89999,9.

1 See text for treatment of motor spirit (SITC 3321) and passenger motor cars (SITC 7321).

as reporters. These countries report trade till 1987. Trade in capital goods largely takes place between the developing country and the group of developed market economies, so that a very high proportion of actual trade should be covered. But this need not hold for intermediate goods as well, since South-South or South-OPEC trade in primary products (especially oil) can be substantial. It is possible to obtain total intermediate imports (sum of energy and non-energy) for those countries that reported trade data to *Comtrade* till 1987 (Argentina, Brazil, Columbia, Malaysia, Uruguay). For the other countries, trade flows have to be reversed and only exports of DMECs to the developing countries are taken into account. As can be seen in Table 2.1, the share of energy products in total intermediate imports for these economies (Chile, Ecuador, Indonesia, Malaysia, Peru, the Philippines, and Venezuela) is relatively small.

Imported industrial raw materials are composed of the SITC lines reported in Table C.3. These are generally all items classified as primary food and beverages, primary industrial supplies, and primary fuels and lubricants by United Nations (1971), excluding the SITC headings that conflict with the matching of the International Statistics Industrial Classification and the Standard International Trade Classification as reported in Table C.1.[4]

Table C.3: Definition of primary industrial supplies

	Standard international trade classification (Revision 1)
Primary industrial supplies	041, 0421, 043, 044, 0451, 0452, 0459, 05462, 0548, 0711, 0721, 121, 21, 2211, 2212, 2213, 2214, 2215, 2216, 2217, 2218, 2311, 2314, 2411, 242, 24401, 2611, 2612, 2621, 2622, 2623, 2625, 2629, 2631, 2632, 2633, 264, 265, 267, 27, 28, 291, 2921, 2922, 2923, 2924, 2925, 2926, 2929, 3214, 3216, 3217, 33101, 3411, 6671, 6672, 6673, 7358

4 The definition of primary industrial supplies is used in Table 2.5.

C.3. National income accounting statistics

National income statistics are obtained from the World Tables of the World Bank (1989d). If necessary, gross domestic product and gross domestic investment data in constant 1980 prices are converted into constant 1980 dollar values with the 1980 nominal exchange rate (IMF 1990c).

The dollar price index of GDP (1980 = 1) and the dollar price index of gross domestic investments (1980 = 1) are computed by converting the implicit price in national currencies (nominal values devided by values in 1980 prices) into the dollar price with the nominal exchange rate and normalizing it to one in the base year.

C.4. Specific variables used in Chapter 5[5]

C.4.1. Implicit average effective import taxes

Implicit average effective import taxes are defined as the ratio of government revenues from import taxes over the nominal import bill. Data on nominal government revenues from imports in national currency stem from IMF (1990b). They are converted into current US dollar figures with the nominal exchange rate.

The structure of charge rates (1987/1988) originates from the Tariff Information System (TIS) of UNCTAD. Capital, intermediate, and consumer goods are defined in the same way as the import categories. Each charge rate is weighted with the import composition of all developing countries in order to mitigate the effect that high charges reduce imports.

The IMF statistics on government revenue from imports relate to total imports. Trade data for countries not reporting flows till 1987 (Chile, Ecuador, Indonesia, Mexico, Peru, the Philippines, and Venezuela) have to be interpolated in order to be able to apply the structure of charges to trade flows in the entire period. For the years in which no import data are supplied by the respective developing country, the trade flows read in from the United Nations *Comtrade* database are reversed, using the developed market economies as the reporter for their exports to each country. South-South and South-OPEC trade takes place mainly in primary intermediate goods. The difference between imports from the DMECs and total imports of the respective developing country (IMF 1990a) is used as a proxy for the trade in primary intermediate imports of the developing country with other developing countries and OPEC.

---

5 For a definition of other variables see the description in Chapter 5 (e.g., the derivation of the protection proxy in Section 5.2.2. and the method of capital stock computation in Section 5.3.2.1.).

C.4.2. Real import capacity

Nominal import capacity is defined as the exports of goods plus balance on services plus balance on government and private transfers plus net long term capital inflow plus errors and omissions (UNCTAD 1988a, p.12, fn a). All nominal data are taken from IMF (1990a). The nominal value is then deflated with the dollar price index of imported capital and, alternatively, with the price index for imported intermediate products. The price indices stem from the aggregation and deflation of trade data (Section C.2.).

C.4.3. Interest rates

Domestic and foreign interest rates are used in the 'modification 2 estimations' in Chapter 5, when the alternative user cost concept is employed.

Nominal dollar interest rates on foreign loans (1972 to 1987) are 'average terms of new commitments on all creditors' from the World Bank (1989b, 1989c).

Domestic interest rates: Sources and definitions differ from country to country:

Columbia: 'Effective nominal interest rate' for the years 1980 to 1987 stem from Economic Comission for Latin America and the Carribean (1989). The figures are interpolated backwards till 1972 using the discount rate from IMF (1990c).

Ecuador: The 'short-term preferential lending rate' from 1980 to 1987 is taken from Economic Commission for Latin America and the Carribean (1989). The backward interpolation corresponds to the Columbian data.

Mexico: The 'lending rate' for the years 1978 to 1984 originates from IMF (1990c). They were interpolated with the 'average cost of funds' series (IMF 1990c).

Philippines: The 'lending rate' is available from 1976 to 1987 (IMF 1990c); it is interpolated backwards with the discount rate from the same source.

Venezuela: The domestic lending rate for the 1972 to 1986 period stems from World Bank (1988a). It is interpolated for 1987 with the lending rate from IMF (1990c).

C.5. Variable definitions and sources, Chapter 6

D_M Dummy variable for the Ecuadorian 'other import' equation.

D_{PX} Dummy variable for the Ecuadorian price equation.

E Nominal exchange rate index (1980 = 1), national currency per US dollar (IMF 1990c).

E_{80} Nominal exchange rate of the base year (1980), national currency per US dollar (IMF 1990c).

J Efficiency capital stock.

K_d Capital stock consisting of domestically produced investment goods. Calculated by inventory method (depreciation rate 12.5%) from real gross domestic investment figures (net of imported capital goods) in constant 1980 prices, national currency (World Bank 1989d). Compare Section 5.3.5.1. and 5.3.5.2.

K_f Capital stock consisting of imported machinery and equipment. Calculated by inventory method (depreciation rate 12.5%) from the series on imported capital goods (M_{Cap}).

L Labor.

M_{Cap} Real imports of capital goods (in 1980 national prices), UN Statistical Office (*Comtrade*) and UNCTAD Database on Price Deflators. See definition in terms of Standard International Trade Classification in Table C.2.[6]

M_{Int} Real imports of intermediate goods (in 1980 national prices), UN Statistical Office (*Comtrade*) and UNCTAD Database on Price Deflators. See definition in terms of Standard International Trade Classification in Table C.2.[7]

M_{Oth} Real imports of other goods (in 1980 national prices). It is calculated as the residual from equation 6.21.

MV Nominal import value (US $). For Columbia from UN Statistical Office (*Comtrade*); for Ecuador from IMF (1990a) since Ecuador did not report trade flows to *Comtrade* till 1987.

P_J Price deflator of the efficiency capital stock.

---

6 Analogously to the estimations in Chapter 5, only the trade flows between the developed market economies and Ecuador and Columbia are taken into account.

7 For Ecuador only the exports of intermediate goods from the developed market economies are taken into account.

P'_{MCap}	Price deflator of 'capital goods imports' (1980=1), US $, UN Statistical Office (*Comtrade*) and UNCTAD Database on Price Deflators. The definition of capital goods follows the classification in Table C.2.
P'_{MInt}	Price deflator of 'intermediate goods imports' (1980=1), US $. UN Statistical Office (*Comtrade*) and UNCTAD Database on Price Deflators. The definition of intermediate goods follows the classification in Table C.2.
P'_{MOth}	Price deflator of 'other imports' (1980=1), US $. Approximated here by the price deflator for consumer good imports. UN Statistical Office (*Comtrade*) and UNCTAD Database on Price Deflators. See classification in Table C.2 above.
P_Q	Gross output deflator (1980=1), national currency. Proxied by the implicit gross output deflator of the industries 381 to 385, United Nations Industrial Development Organization Data Base (Global Studies).
P_W	Price deflator of OECD income (1980=1), US $ (OECD 1989).
P_X	Price deflator of exports (1980=1), national currency. UNCTAD Database on Price Deflators.
P_Y	Gross domestic product deflator (1980=1), national currency (World Bank 1989d).
Q	Real gross domestic output.
t	Time trend for technological progress.
TB	Trade balance in current US $.
t_{Int}	Implicit average import tax on intermediate goods, UNCTAD Tariff Information System. See Section C.4.1.
t_{Cap}	Implicit average import tax on capital goods, UNCTAD Tariff Information System. See Section C.4.1.
t_{Oth}	Implicit average import tax on 'other imports', approximated by the calculated tariff rate on imports of consumer goods, UNCTAD Tariff Information System. See Section C.4.1.
w	Nominal wage rate index of manufacturing (1980=100), national currency. UNIDO Data Base (Global Studies).

X Real exports in constant 1980 national currencies. UN Statistical Office (Comtrade) and UNCTAD Database on Price Deflators for Columbia; IMF (1990a) and UNCTAD Database on Price Deflators for Ecuador.

XV Nominal export value (US $). For Columbia from UN Statistical Office (*Comtrade*); for Ecuador from IMF (1990a) since Ecuador did not report trade flows till 1987.

Y Real Gross Domestic Product (in constant 1980 national currency), World Bank (1989d).

Y_W Real Gross Domestic Product of the OECD countries in 1980 dollars (OECD 1989). Converted into national currency of Ecuador and Columbia with the nominal exchange rate (IMF 1990c).

C.6. Computational notes

All aggregation and deflation of raw data was conducted while visiting the United Nations Conference on Trade and Development (Geneva) and the United Nations Industrial Development Organization (Vienna). All other computations of variables, estimations, and simulations presented in this study were carried out with the statistical program *SAS* on the BASF mainframe computer at the University of Konstanz. Non-linear estimations and simultaneous non-linear simulations employed the Gauss-Newton method.

REFERENCES

Amelung, T., and F.L. Sell (1991), Tariff Redundancy and Its Implication for Protection Substitution, Konjunkturpolitik 37, pp.66-86.

Appelbaum, E., and U.R. Kohli (1979), Canada-United States Trade: Tests for the Small-Open-Economy Hypothesis, Canadian Journal of Economics 12, pp.1-14.

Arize, A. (1990), An Econometric Investigation of Export Behaviour in Seven Asian Developing Countries, Applied Economics 22, pp.891-904.

Armington, P.S. (1969), A Theory of Demand for Products Distinguished by Place of Production, IMF Staff Papers 16, pp.159-178.

Arrieta, G.M.G. (1988), Interest Rates, Savings, and Growth in LDCs: An Assessment of Recent Empirical Research, World Development 16, pp.589-605.

Arrow, K.J., H.B. Chenery, B.S. Minhas, and R.M. Solow (1961), Capital-Labor Substitution and Economic Efficiency, Review of Economics and Statistics 43, pp. 225-250.

Aw, B.Y., and M.J. Roberts (1985), The Role of Imports from the Newly-Industrializing Countries in U.S. Production, Review of Economics and Statistics 67, pp.108-117.

Bacha, E.L. (1986), Terms of Reference for the Country Studies, World Development 14, pp.909-918.

Balassa, B. (1984), Adjustment Policies in Developing Countries: A Reassessment, World Development 12, pp.955-972.

Ballance, R., and S. Sinclair (1983), Collapse and Survival: Industry Strategies in a Changing World, London: Allen and Unwin.

Bardhan, P.K. (1965), Equilibrium Growth in the International Economy, Quaterly Journal of Economics 79, pp.455-464.

Bardhan, P.K., and S. Lewis (1970), Models of Growth with Imported Inputs, Economica 37, pp.373-385.

Bautista, R.M. (1980), Import Demand for Capital Equipment in the Philippines, Weltwirtschaftliches Archiv 116, pp.560-573.

Bautista, R.M., H. Hughes, D. Linn, D. Morawetz, and F.E. Thoumi (1981), Capital Utilization in Manufacturing, New York et al.: Oxford University Press.

Behrman, J.R. (1971), Review Article: UNCTAD Secretariat, Trade Prospects and Capital Needs of Developing Countries, International Economic Review 12, pp.519-525.

Bergsten, C.F., W.R. Cline, and J. Williamson (1985), Bank Lending to Developing Countries: The Policy Alternatives, Policy Analyses in International Economics 10, Institute for International Economics, Washington.

Berndt, E.R., and L.R. Christensen (1973), The Internal Structure of Functional Relationships: Separability, Substitution, and Aggregation, Review of Economic Studies 40, pp.403-410.

Berndt, E.R., and D.O. Wood (1975), Technology, Prices, and the Derived Demand for Energy, Review of Economics and Statistics 57, pp.259-268.

Bhagwati, J.N. (1978), Foreign Trade Regimes and Economic Development: Anatomy and Consequences of Exchange Control Regimes, Cambridge: Ballinger.

Bhagwati, J.N. (1988), Export-Promoting Trade Strategy: Issues and Evidence, The World Bank Research Observer 3, pp.27-57.

Bhattacharya, A., and J.F. Linn (1988), Trade and Industrial Policies in the Developing Countries of East Asia, World Bank Discussion Papers No. 27, Washington.

Bocco, A.M. (1990), Ecuador: The State, Public Policy and Capital Accumulation, 1970-1983, in: C. Anglade and C. Fortin (eds.), The State and Capital Accumulation in Latin America (2), London: MacMillan, pp.88-105.

Bond, M.E. (1987), An Econometric Study of Primary Commodity Exports from Developing Regions to the World, IMF Staff Papers 34, pp.191-227.

Bruno, M., and J. Sachs (1985), Economics of Worldwide Stagflation, Cambridge: Harvard University Press.

Buffie, E.F. (1986), Devaluation, Investment and Growth in LDCs, Journal of Development Economics 20, pp.361-379.

Buffie, E.F. (1989), Mexico 1958-1986: From Stabilizing Development to the Debt Crisis, in: J.D. Sachs (ed.), Developing Country Debt and the World Economy, Chicago: Chicago University Press, pp.141-168.

Burgress, D.F. (1974a), A Cost-Minimization Approach to Import Demand Equations, Review of Economics and Statistics 56, pp.225-234.

Burgress, D.F. (1974b), Production Theory and the Derived Demand for Imports, Journal of International Economics 4, pp.103-117.

Cardoso, E., and A. Fishlow (1989a), Latin American Economic Development: 1950-1980, NBER Working Paper No. 3161, Cambridge.

Cardoso, E., and A. Fishlow (1989b), The Macroeconomics of the Brazilian Debt, in: J.D. Sachs (ed.), Developing Country Debt and the World Economy, Chicago: Chicago University Press, pp.81-99.

Chenery, H.B., and M. Bruno (1962), Development Alternatives in an Open Economy: The Case of Israel, Economic Journal 72, pp.79-103.

Chenery, H.B., S. Robinson, and M. Syrquin (1986), Industrialization and Growth, New York et al.: Oxford University Press.

Chenery, H.B., and A.M. Strout (1966), Foreign Assistance and Economic Development, American Economic Review 56, pp.679-733.

Christensen, L.R., D.W. Jorgensen, and L.J. Lau (1971), Conjugate Duality and the Transcendental Logarithmic Function, Econometrica 39 (Supplement), pp.255-256.

Chudnovsky, D., M. Nagoa, and S. Jacobsson (1983), Capital Goods Production in the Third World, London: Frances Pinter.

Cline, W.R. (1983), International Debt and the Stability of the World Economy, Policy Analyses in International Economics 4, Institute for International Economics, Washington.

Cline, W.R. (1984), International Debt: Systematic Risk and Policy Response, Institute for International Economics, Washington.

Coen, R. M. (1971), The Effect of Cash Flow on the Speed of Adjustment, in: G.Fromm (ed.), Tax Incentives and Capital Spending, Brookings Institution, Washington, pp.131- 196.

Coen, R.M., and B.G. Hickman (1970), Constrained Joint Estimation of Factor Demand and Production Functions, Review of Economics and Statistics 52, pp.287-300.

Collins, S.M. (1988), South Korea's Experience with External Debt, NBER Working Paper No. 2598, Cambridge.

Collins, S.M., and W.-A. Park (1989), External Debt and Macroeconomic Performance in South Korea, in: J.D. Sachs (ed.), Developing Country Debt and the World Economy, Chicago: Chicago University Press, pp.121-140.

Conway, P. (1986), Decomposing the Determinants of Trade Deficits: Turkey in the 1970s, Journal of Development Economics 21, pp.235-258.

Corsepius, U. (1989), Peru at the Brink of Economic Collapse: Current Problems and Policy Options, Kiel Discussion Papers No. 153, Kiel.

Corsepius, U., and A. Schipke (1989), Die Computerindustrie in Schwellenländern - der Fall Brasilien, Die Weltwirtschaft (1), pp.137-151.

Costrell, R.M., G.E. Duguay, and G.I. Treyz (1986), Labour Substitution and Complementarity Among Age-Sex Groups, Applied Economics 18, pp.777-791.

Currie, D., A. Muscatelli, and D. Vines (1988), Introduction, in: D. Currie and D. Vines (eds.), Macroeconomic Interactions between North and South, Cambridge et al.: Cambridge University Press, pp.1-31.

Dadkhah, K.M., and F. Zahedi (1986), Simultaneous Estimation of Production Functions and Capital Stocks for Developing Countries, Review of Economics and Statistics 68, pp.443-451.

Dahlmann, C.J., B. Ross-Larson, and L.E. Westphal (1987), Managing Technological Development: Lessons from the Newly Industrializing Countries, World Development 15, pp.759-775.

Dervis, K., J. de Melo, and S. Robinson (1981), A General Equilibrium Analysis of Foreign Exchange Shortages in a Developing Economy, World Bank Staff Working Paper No. 443, Washington.

Dervis, K., J. de Melo, and S. Robinson (1982), General Equilibrium Models for Development Policy, Cambridge et al.: Cambridge University Press.

Dick, H., E. Gerken, and D.P. Vincent (1982a), The Benefits of the CAP for Developing Countries: A Case Study of the Ivory Coast, European Review of Agricultural Economics 9, pp.157-181.

Dick, H., E. Gerken, T. Mayer, and D.P. Vincent (1982b), Stabilisation Strategies in Primary Commodity Exporting Countries: a Case Study of Chile, Kiel Working Papers No. 144, Kiel.

Dohner, R.S., and P. Intal (1989), Debt Crisis and Adjustment in the Philippines, in: J.D. Sachs (ed.), Developing Country Debt and the World Economy, Chicago: Chicago University Press, pp. 169-191.

Donges, J.B. (1976), A Comparative Survey of Industrialization Policies in Fifteen Semi-Industrial Countries, Weltwirtschaftliches Archiv 112, pp.626-659.

Donges, J.B., and J. Riedel (1977), The Expansion of Manufactured Exports in Developing Countries: An Empirical Assessment of Supply and Demand Issues, Weltwirtschaftliches Archiv 113, pp.58-87.

Dooley, M., W. Helkie, R. Tryon, and J. Underwood (1986), An Analysis of External Debt Positions of Eight Developing Countries through 1990, Journal of Development Economics 21, pp.283-318.

Dornbusch, R. (1985), Policy and Performance Links between LDC Debtors and Industrial Nations, Brookings Papers on Economic Activity (2), pp.303-356.

Dornbusch, R., and Y.C. Park (1987), Korean Growth Policy, Brookings Papers on Economic Activity (2), pp.389-454.

Durbin, J. (1970), Testing for Serial Correlation in Least Squares Regression when Some of the Regressors Are Lagged Dependent Variables, Econometrica 38, pp.410-421.

Dutta, M. (1964), A Prototype Model of India's Foreign Sector, International Economic Review 5, pp.82-103.

Ebenroth, C.T., and H. Jäger (1990), Die Entwicklung Mexikos im Zeichen der Brady-Initiative, Aussenwirtschaft 45, pp.237-265.

Economic Commission for Latin America and the Carribean (1989), Economic Survey of Latin America 1987, Santiago.

Edwards, S. (1989), Openness, Outward Orientation, Trade Liberalization and Economic Performance in Developing Countries, NBER Working Paper No. 2908, Cambridge.

Eggers, G. (1988), Chile, Grundlagen und Grundzüge der Wirtschaftspolitik, Studien der Bremer Gesellschaft zur Wirtschaftsforschung e.V., Bremen.

Eggerstedt, H., J. Hentschel, and T.J. Noh (1987), Quantitative Trade Restrictions Against Developing Countries, Asian Economies 62, pp.5-23.

El-Shibly, M., and A.P. Thirlwall (1981), Dual-Gap Analysis for the Sudan, World Development 9, pp.193-200.

Erbe, R. (1982), Foreign Indebtedness and Economic Growth: The Philippines, Intereconomics 17, pp.125-132.

Faini, R., L. Pritchett, and F. Clavijo (1988), Import Demand in Developing Countries, PPR Working Paper Series Trade Policy No. 122, World Bank, Washington.

Fajana, O. (1979), Trade and Growth: The Nigerian Experience, World Development 7, pp.73-78.

Farzin, Y.H. (1988), The Relationship of External Debt and Growth, World Bank Discussion Papers No. 24, Washington.

Feder, G. (1981), Growth and External Borrowing in Trade Gap Economies of Less Developed Countries, Aussenwirtschaft 36, pp.381-395.

Feltenstein, A., D. Lebow, and S. van Wijnbergen (1990), Savings, Commodity Market Rationing, and the Real Rate of Interest in China, Journal of Money, Credit and Banking 22, pp.234-252.

Field, A.J. (1987), An Estimate of the Textile and Clothing Sector Production Function for Selected Countries in the Early 1970s, World Development 15, pp.1285-1290.

Findlay, R., and C.A. Rodriguez (1977), Intermediate Imports and Macroeconomic Policy Under Flexible Exchange Rates, Canadian Journal of Economics 10, pp.208-217.

Fisher, F.M. (1982), Aggregate Production Functions Revisted: The Mobility of Capital and the Rigidity of Thought, Review of Economic Studies 49, pp.615-626.

Fitchett, D.A. (1976), Capital-Labor Substitution in the Manufacturing Sector of Panama, Economic Development and Cultural Change 24, pp.577-592.

Foss, M.F. (1963), The Utilization of Capital Equipment: Postwar Compared with Prewar, Survey of Current Business 43, pp.8-16.

Fritsch, W. (1988), Brazil's Growth Prospects: Domestic Savings, External Finance and OECD Performance Interactions, in: D. Currie and D. Vines (eds.), Macroeconomic Interactions between North and South, Cambridge et al.: Cambridge University Press, pp.253-278.

Fuss, M., D. McFadden, and Y. Mundlak (1978), A Survey of Fundamental Forms in the Economic Analysis of Production, in: M. Fuss and D. McFadden (eds.), Production Economics: A Dual Approach to Theory and Applications, Amsterdam: North-Holland, pp.219-268.

Gallant, A.R. (1987), Nonlinear Statistical Models, New York et al.: John Wiley & Sons.

Ghose, D., A.K. Lahiri, and W. Wadhwa (1986), Quantitative Restrictions and Indian Imports, Journal of Development Economics 20, pp.247-262.

Gilbert, C.L. (1986), The Impact of Exchange Rates and Developing Country Debt on Commodity Prices, Division Working Paper 1986-4, World Bank Commodity Studies and Projections Division, Washington.

Glick, R., and H.J. Kharas (1986), The Costs and Benefits of Foreign Borrowing: A Survey of the Multi-Period Models, Journal of Development Studies 22, pp.279-299.

Goldsbrough, D., and I. Zaidi (1986), Transmission of Economic Influences from Industrial to Developing Countries, IMF Staff Studies for the World Economic Outlook (July), pp.150-195.

Goldstein, M., and M.S. Khan (1978), The Supply and Demand for Exports: A Simultaneous Approach, Review of Economics and Statistics 60, pp.275-286.

Goldstein, M., and M.S. Khan (1985), Income and Price Effects in Foreign Trade, in: R.W. Jones and P.B. Kenen (eds.), Handbook of International Economics (2), Amsterdam: North-Holland, pp.1041-1105.

Greenaway, D. (1983), International Trade Policy: From Tariffs to the New Protectionism, London: MacMillan.

Hallett, A.H. (1988), Commodities, Debt and North-South Cooperation: a Cautionary Tale from the Structuralist Camp, in: D. Currie and D. Vines (eds.), Macroeconomic Interactions between North and South, Cambridge et al.: Cambridge University Press, pp.106-140.

Hamermesh, D.S., and J. Grant (1979), Econometric Studies of Labor-Labor Substitution and their Implications for Policy, Journal of Human Resources 14, pp.518-542.

Hamilton, C. (1984), Voluntary Export Restraints: ASEAN Systems for Allocation of Export Licences, Seminar Paper No. 275, Institute for International Economic Studies, Stockholm.

Heathfield, D.F. (1972), The Measurement of Capital Usage Using Electricity Consumption Data for the UK, Journal of the Royal Statistical Society 135, pp.208-220.

Hemphill, W.L. (1974), The Effect of Foreign Exchange Receipts on Imports of Less Developed Countries, IMF Staff Papers 21, pp.637-677.

Hentschel, J. (1988), Non-Optimizing Growth-Cum-Debt Models and Sustainability of Indebtedness, The Indian Journal of Economics 273, pp.147-161.

Hentschel, J. (1989), The Changing Composition of Imports of Highly Indebted Countries, Economia Internazionale 42, pp.79-89.

Herberg, H., H. Hesse, and A. Schuseil (1982), Imports of Intermediate Goods and the Efficacy of Fiscal Policy under Flexible Exchange Rates, Weltwirtschaftliches Archiv 118, pp.104-130.

Herrmann, R. (1988), Internationale Agrarmarktabkommen: Analyse ihrer Wirkungen auf den Märkten für Kaffee und Kakao, Kieler Studien 215, Tübingen: J.C.B.Mohr.

Hicks, N.L. (1976), A Model of Trade and Growth for the Developing World, European Economic Review 7, pp.239-255.

Hirsch, S. (1974), Capital or Technology? Confronting the Neo-Factor Proportions and Neo-Technology Accounts of International Trade, Weltwirtschaftliches Archiv 110, pp.535-563.

Hirsch, S. (1975), The Product Cycle Model or International Trade - A Multi-Country Cross-Section Analysis, Oxford Bulletin of Economics and Statistics 37, pp.305-317.

Hock, L.K. (1986), Malaysia: The Structure and Causes of Manufacturing Sector Protection, in: C. Findlay and R. Garnaut (eds.), The Political Economy of Manufacturing Protection: Experiences of ASEAN and Australia, Sydney et al.: Allen and Unwin, pp.99-134.

Hoffmann, L., and B. Weber (1976), Economies of Scale, Factor Intensities and Substitution: Micro Estimates for Malaysia's Manufacturing Industries, Weltwirtschaftliches Archiv 112, pp.111-135.

Houthakker, H.S., and S.P. Magee (1969), Income and Price Elasticities in World Trade, Review of Economics and Statistics 51, pp.111-125.

Ikemoto, Y. (1986), Technical Progress and Level of Technology in Asian Countries, 1970-1980: A Translog Index Approach, The Developing Economies 24, pp.368-390.

Inter-American Development Bank (1984), External Debt and Economic Development in Latin America, Washington.

International Monetary Fund (1983), Interest Rate Policies in Developing Countries, Occasional Papers No. 22, Washington.

International Monetary Fund (1989), World Economic Outlook (April), Washington.

International Monetary Fund (1990a), Balance of Payments Statistics, on tape, Washington.

International Monetary Fund (1990b), Government Finance Statistics, on tape, Washington.

International Monetary Fund (1990c), International Financial Statistics, on tape, Washington.

Intriligator, M.D. (1978), Econometric Models, Techniques and Applications, Englewood Cliffs: Prentice-Hall.

Itsumi, Y., and C. Morigushi (1978), Terms of Trade and Full Capacity Growth in a Resource-Importing Economy, Journal of International Economics 8, pp.117-126.

Jacobsson, S. (1988), Intra-Industry Specialization and Development Models for the Capital Goods Sector, Weltwirtschaftliches Archiv 124, pp.14-37.

Johnson, H.G. (1953), Equilibrium Growth in an International Economy, Canadian Journal of Economics and Political Science 19, pp.478-500.

Johnson, H.G. (1958), International Trade and Economic Growth, Studies in Pure Theory, London: Allen and Unwin.

Johnston, J. (1984), Econometric Methods, Singapore et al.: McGraw-Hill.

Jones, R.W. and D. Purvis (1983), International Differences in Response to Common External Shocks: The Role of Purchasing Power Parity, in: E. Claasen and P. Salin (eds.), Recent Issues in the Theory of Flexible Exchange Rates, Amsterdam: North-Holland, pp.33-55.

Kaneda, H. (1982), Specification of Production Functions for Analyzing Technical Change and Factor Inputs in Agricultural Development, Journal of Development Economics 11, pp.97-108.

Katrak, H. (1985), Imported Technology, Enterprice Size and R&D in a Newly Industrializing Country: The Indian Experience, Oxford Bulletin of Economics and Statistics 47, pp.213-229.

Katz, J.M. (1969), Production Functions, Foreign Investment and Growth, Amsterdam: North-Holland.

Kell, G. and S. Marchese (1991), Exports of Textiles and Metals: The Question of Sustainability and Reasoned Growth, UNCTAD Review, forthcoming.

Khan, M. S. (1974), Import and Export Demand in Developing Countries, IMF Staff Papers 21, pp.678-693.

Khan, M.S. (1975), The Structure and Behavior of Imports of Venezuela, Review of Economics and Statistics 57, pp.221-224.

Khan, M.S., and M.D. Knight (1986), Import Compression and Export Performance in Developing Countries, Discussion Paper No. 197, Economics and Research Staff, Development Research Department, World Bank, Washington.

Khan, M.S., and M.D. Knight (1988), Import Compression and Export Performance in Developing Countries, Review of Economics and Statistics 70, pp.315-321.

Khatkhate, D.R. (1988), Assessing the Impact of Interest Rates in Less Developed Countries, World Development 16, pp.577-588.

Kirkpatrick, C.H., N. Lee, and F.I. Nixson (1984), Industrial Structure and Policy in Less Developed Countries, London: Allen and Unwin.

Kohli, U.R. (1978), A Gross National Product Function and the Derived Demand for Imports and Supply of Exports, Canadian Journal of Economics 11, pp.167-182.

König, P. (1990), Ölpreisschocks und Makroökonomische Anpassungsprozesse zwischen Industrieländern, Konstanz: Hartung-Gorre.

Koutsoyiannis, A. (1977), Theory of Econometrics, London et al.: MacMillan.

Krueger, A.O. (1978), Foreign Trade Regimes and Economic Development: Liberalization Attempts and Consequences, Cambridge: Ballinger.

Krueger, A.O. (1985), Import Substitution verus Export Promotion, Finance and Development 22, pp.20-23.

Kürzinger, E. (1988), Argentinien: Die blockierte Entwicklung, Deutsches Institut für Entwicklungspolitik, Berlin.

Laird, S., and J. Nogués (1988a), Manufactured Export Performance of the Highly Indebted Countries, The Developing Economies 26, pp.403-421.

Laird, S., and J. Nogués (1988b), Trade Policies and the Debt Crisis, PPR Working Papers Series International Trade No. 99, World Bank, Washington.

Langhammer, R.J. (1988), Tariff Reductions and Tariff Redundancy in ASEAN Countries, ASEAN Economic Bulletin 4, pp.252-270.

Laumas, P.S., and M. Williams (1981), The Elasticity of Substitution in India's Manufacturing Sector, Journal of Development Economics 8, pp.325-337.

Leven, R., and D.L. Roberts (1983), Latin America's Prospects for Recovery, Federal Reserve Bank of New York Quaterly Review 8 (3), pp.6-13.

Little, I., T. Scitovsky, and M. Scott (1970), Industry and Trade in Some Developing Countries, London et al.: Oxford University Press.

Lopez, R., and V. Thomas (1988), Imports and Growth in Africa, PPR Working Paper Series Trade Policy No. 20, World Bank, Washington.

Lucas, R.E.B. (1989), Liberalization of Indian Trade and Industrial Licensing, Journal of Development Economics 31, pp.141-175.

Marques, J.R. (1984), Oil Price Effects and OPEC's Pricing Policy, Lexington: LexingtonBooks.

Marquez, J.R. (1985), Foreign Exchange Constraints and Growth Possibilities in the LDCs, Journal of Development Economics 19, pp.39-57.

Marquez, J.R., and P. Pauly (1987), International Policy Coordination and Growth Prospects of Developing Countries, Journal of Development Economics 25, pp.89-104.

McCarthy, F.D., L. Taylor, and C. Talati (1987), Trade Patterns in Developing Countries, 1964-1982, Journal of Development Economics 27, pp.5-39.

McKinnon, R.I. (1964), Foreign Exchange Constraints in Economic Development and Efficient Aid Allocation, Economic Journal 74, pp.388-409.

Melo, J. de, and S. Robinson (1989), Product Differentiation and the Treatment of Foreign Trade in Computable General Equilibrium Models of Small Economies, Journal of International Economics 27, pp.47-67.

Melo, O., and M.G. Vogt (1984), Determinants of the Demand for Imports of Venezuela, Journal of Development Economics 14, pp.351-358.

Michaely, M. (1984), Trade, Income Levels and Dependence, Studies in International Economics 8, Amsterdam: North-Holland.

Michalopoulos, C. (1975), Production and Substitution in Two-Gap Models, Journal of Development Studies 11, pp.343-356.

Mirakhor, A., and P. Montiel (1987), Import Intensity of Output Growth in Developing Countries 1970-1985, IMF Staff Studies for the World Economic Outlook (August), pp.59-97.

Mohabbat, K.A., and A.J. Desai (1983), Factor Substitution and Import Demand for South Korea: A Translog Analysis, Weltwirtschaftliches Archiv 119, pp.709-723.

Mohabbat, K.A., A.J. Desai, and M. Williams (1984), Import Demand for India: A Translog Cost Function Approach, Economic Development and Cultural Change 32, pp.593-605.

Moran, C. (1989), Imports under a Foreign Exchange Constraint, The World Bank Economic Review 3, pp.279-295.

Morawetz, D. (1976), Elasticities of Substitution in Industry: What Do We Learn from Econometric Estimates?, World Development 4, pp.11-15.

Mukerji, V. (1963), A Generalized SMAC Function with Constant Elasticities of Substitution, Review of Economic Studies 30, pp.233-236.

Mulate, T. (1980), Capital-Labor Substitution in the Ethiopian Manufacturing Industries, The Developing Economies 18, pp.275-287.

Nerlove, M. (1967), Recent Empirical Studies of the CES and Related Production Functions, in: M. Brown (ed.), The Theory and Empirical Analysis of Production, New York et al.: Columbia University Press, pp.55-122.

Nunnenkamp, P. (1983), Rising Oil Prices and Industrial Production in Some Developing Countries in the Seventies, Kiel Working Papers No. 190, Kiel.

Oberhänsli, H. (1981), Note: Wachstum und Aussenverschuldung - Zwei Projektionen zum Trade-Gap Modell von Feder, Aussenwirtschaft 36, pp.396-402.

Obstfeld, M. (1980), Intermediate Imports, the Terms of Trade, and the Dynamics of the Exchange Rate and Current Account, Journal of International Economics 10, pp.461-480.

Oniki, H., and H. Uzawa (1965), Patterns of Trade and Investment in a Dynamic Model of International Trade, Review of Economic Studies 32, pp.15-38.

Organisation for Economic Co-Operation and Development (1989), National Accounts 1960-1987, Paris.

Park, S.-H. (1985), Investment Planning and the Macroeconomic Constraints in Developing Countries: The Case of the Syrian Arab Republic, World Development 13, pp.837-853.

Pindyck, R.S., and J.J. Rotemberg (1983a), Dynamic Factor Demands and the Effects of Energy Price Shocks, American Economic Review 73, pp.1066-1079.

Pindyck, R.S., and J.J. Rotemberg (1983b), Dynamic Factor Demands under Rational Expectations, Scandinavian Journal of Economics 85, pp.223-238.

Pindyck, R.S., and D.L. Rubinfeld (1981), Econometric Models and Economic Forecasts, Singapore et al.: McGraw-Hill.

Prais, S.J., and C.B. Winsten (1954), Trend Estimators and Serial Correlation, Cowles Commission Discussion Papers No. 383, Chicago.

Ranis, G. (1979), Appropriate Technology: Obstacles and Opportunities, in: S.M. Rosenblatt (ed.), Technology and Economic Development: A Realistic Perspective, Boulder: Westview Press, pp.23-58.

Rashid, M.A. (1984), Modelling the Developing Economy: The Philippines, The Developing Economies 22, pp.274-288.

Rauscher, M. (1989), OPEC and the Price of Oil, Berlin et al.: Springer.

Richtering, J. (1989), The Debt Situation of the Developing Countries: A Medium-Term Scenario Analysis, in: W. Krelle (ed.), The Future of the World Economy, Berlin et al.: Springer, pp.527-539.

Roemer, M. (1981), Dependence and Industrialization Strategies, World Development 9, pp.429-434.

Ronning, G. (1991), Hierarchical Import Allocation and Institutional Arrangements: An Econometric Analysis of the West-German Coffee Imports, Empirical Economics, forthcoming.

Sachs, J.D. (1987), Trade and Exchange Rate Policies in Growth-Oriented Adjustment Programs, NBER Working Paper No. 2226, Cambridge.

Sachs, J.D. (1989), Introduction, in: J.D. Sachs (ed.), Developing Country Debt and the World Economy, Chicago: Chicago University Press, pp. 1-33.

Salas, J. (1982), Estimation of the Structure and Elasticities of Mexican Imports in the Period 1961-1979, Journal of Development Economics 10, pp.297-311.

Sankar, U. (1970), Elasticities of Substitution and Returns to Scale in Indian Manufacturing Industries, International Economic Review 11, pp.399-411.

Sanyal, K.K. and R.W. Jones (1982), The Theory of Trade in Middle Products, American Economic Review 72, pp.16-31.

Sarmad, K. (1989), The Determinants of Import Demand in Pakistan, World Development 17, pp.1619-1625.

Sato, K. (1967), A Two-Level Constant-Elasticity-of-Substitution Production Function, Review of Economic Studies 34, pp.201-218.

Sato, K. (1975), Production Functions and Aggregation, Amsterdam: North-Holland.

Saunders, P., and A. Dean (1986), The International Debt Situation and Linkages between Developing Countries and the OECD, OECD Economic Studies 7, pp.155-203.

Schiavo-Campo, S. (1980), Limited Factor Substitutability, Economic Dualism, and Non-Saving Constraints on Growth, Journal of Development Economics 7, pp.283-294.

Schmid, M. (1982), Stagflationary Effects of a Devaluation in a Monetary Model with Imported Intermediate Inputs, Jahrbücher für Nationalökonomie and Statistik 197, pp.107-129.

Shiells, C.R. (1989), Competition and Complementarity between US Imports from Developed and Newly Industrializing Countries, Weltwirtschaftliches Archiv 125, pp.114-128.

Shiells, C.R., R.M. Stern, and A.V. Deardorff (1986), Estimates of the Elasticities of Substitution between Imports and Home Goods for the United States, Weltwirtschaftliches Archiv 122, pp.497-519.

Siebert, H. (1986), Allokationseffekte und Importbeschränkungen. Eine Anmerkung, Zeitschrift für Wirtschafts- und Sozialwissenschaften 106, pp.509-516.

Siebert, H. (1989), Güterwirtschaftliche Anpassungsprozesse zur Lösung der Verschuldungsfrage, in: G. Bombach, B. Gahlen, and A.E. Ott (eds.), Die Nationale und Internationale Schuldenproblematik, Tübingen: J.C.B.Mohr, pp.73-94.

Siebert, H. (1991), Außenwirtschaft, 5th edition, Stuttgart: Gustav Fischer.

Sines, R.H. (1979), Sectoral Elasticities of Substitution between Labour and Capital in Venezuelan Manufacturing: A Cross-Section Micro Analysis, World Development 7, pp.79-82.

Spanos, A (1986), Statistical Foundations of Econometric Modelling, Cambridge: Cambridge University Press.

Stewart, F. (1977), Technology and Underdevelopment, London: MacMillan.

Strassmann, P.W. (1968), Technological Change and Economic Development, Ithaca: Cornell University Press.

Suttle, P. (1986), Debt Projection Models: A Survey, Oxford Review of Economic Policy 2, pp.62-77.

Takagi, Y. (1981), Aid and Debt Problems in Less Developed Countries, Oxford Economic Papers 33, pp.323-337.

Tan, N.A. (1986), The Philippines: The Structure and Causes of Manufacturing Sector Protection, in: C. Findlay and R. Garnaut (eds.), The Political Economy of Manufacturing Protection: Experiences of ASEAN and Australia, Sydney et al.: Allen and Unwin, pp.48-76.

Taylor, L. (1979), Macro Models for Developing Countries, New York et al.: McGraw-Hill.

Turnovsky, S.J. (1968), International Trading Relationships for a Small Country: The Case of New Zealand, Canadian Journal of Economics 1, pp.772-790.

United Nations (1971), Classification by Broad Economic Categories, Statistical Papers Series M No. 53, Department of Economics and Social Affairs, Statistical Office of the United Nations, New York.

United Nations Conference on Trade and Development (1981), Planning the Technological Transformation of Developing Countries, New York.

United Nations Conference on Trade and Development (1985a), The Capital Goods Sector in Developing Countries: Technology Issues and Policy Options, New York.

United Nations Conference on Trade and Development (1985b), Trade and Development Report 1985, New York.

United Nations Conference on Trade and Development (1986), The UNCTAD Trade Deflator System for Commodity Groupings, Money, Finance and Development Division, Geneva.

United Nations Conference on Trade and Development (1987a), Classification by Broad Technological Categories, Money, Finance and Development Division, Geneva.

United Nations Conference on Trade and Development (1987b), Handbook of Trade Control Measures of Developing Countries, New York.

United Nations Conference on Trade and Development (1987c), Handbook of Trade Control Measures of Developing Countries (Supplement), New York.

United Nations Conference on Trade and Development (1988a), Recent Trends in International Technology Flows and their Implications for Development, Trade and Development Board, Committee on Transfer of Technology, Geneva.

United Nations Conference on Trade and Development (1988b), Trade and Development Report 1988, New York.

United Nations Conference on Trade and Development (1989), Trade and Development Report 1989, New York.

United Nations Conference on Trade and Development (1990), Handbook of Trade and Development Statistics 1989, New York.

United Nations Industrial Development Organization (1974), Industrial Development Survey, Special Issue for the Second General Conference of UNIDO, New York.

United Nations Industrial Development Organization (1979), World Industry Since 1960: Progress and Prospects, Special Issue of the Industrial Development Survey for the Third General Conference of UNIDO, New York.

United Nations Industrial Development Organization (1983), Industry in a Changing World, Special Issue of the Industrial Development Survey for the Fourth General Conference of UNIDO, New York.

United Nations Industrial Development Organization (1984), Conceptual Model for Determining a Country's Desired Level of External Borrowing from the Debtor Country's Perspective, Global and Conceptual Studies Branch, Vienna.

United Nations Industrial Development Organization (1987), Industry and Development, Global Report 1987, Vienna.

United Nations Industrial Development Organization (1988a), Handbook of Industrial Statistics 1988, Vienna.

United Nations Industrial Development Organization (1988b), Industry and Development, Global Report 1988/1989, Vienna.

Uzawa, H. (1962), Production Functions with Constant Elasticities of Substitution, Review of Economic Studies 29, pp.291-299.

Van Bochove, C.A. (1982), Imports and Economic Growth, Boston: Martinus Nijhoff.

Voivodas, C. (1973), Exports, Foreign Capital Inflow and Economic Growth, Journal of International Economics 3, pp.337-349.

Weiss, J. (1988), Industry in Developing Countries, London et al.: Croom Helm.

Weisskopf, R. (1979), Trade Protection and Import Elasticities for Brazil, Review of Economics and Statistics 61, pp.58-66.

Whalley, J. (1989), Recent Trade Liberalization in the Developing World: What is Behind It, and Where Is It Headed?, NBER Working Paper No. 3057, Cambridge.

Williamson, J. (1990), The Progress of Policy Reform in Latin America, Policy Analyses in International Economics 28, Institute for International Economics, Washington.

Winston, G.C. (1974), Factor Substitution, Ex Ante and Ex Post, Journal of Development Economics 1, pp.145-163.

Winters, A. (1987), An Empirical Intertemporal Model of Developing Countries' Imports, Weltwirtschaftliches Archiv 123, pp.58-80.

Woo, W.T., and A. Nasution (1989), The Conduct of Economic Policies in Indonesia and Its Impact on External Debt, in: J.D. Sachs (ed.), Developing Country Debt and the World Economy, Chicago: Chicago University Press, pp. 101-120.

World Bank (1983), Brazil: Industrial Policies and Manufactured Exports, A World Bank Country Study, Washington.

World Bank (1985), Peru. Country Economic Memorandum, Report No. 5806-PE, Washington.

World Bank (1987a), Chile. Adjustment and Recovery, Report No. 6726-CH, Washington.

World Bank (1987b), Columbia. Country Economic Memorandum, Report No. 6626-CO, Washington.

World Bank (1988a), Country Economic Report on Venezuela: Policy Choices and Economic Growth, Report No. 6711-VE, Washington.

World Bank (1988b), World Tables 1988, Washington.

World Bank (1989a), Country Economic Memorandum on Uruguay, Report No. 7424-UR, Washington.

World Bank (1989b), World Bank Debt Tables 1989/1990, Washington.

World Bank (1989c), World Bank Debt Tables 1989/1990 (Supplement), Washington.

World Bank (1989d), World Tables, on tape, Washington.

Zaidi, I.H. (1988), A Rationing Model of Imports and the Balance of Payments in Developing Countries: Theoretical Framework and an Application to the Philippine Economy, Applied Economics 20, pp.43-61.

Ziesemer, T. (1987), Theoretische Ansätze zur Erklärung der Verschuldungssituation der Entwicklungsländer, in: Ökonomie und Gesellschaft 4, Entwicklungsländer und Weltmarkt, Regensburg: Campus, pp.112-156.

Zietz, J., and A. Valdes (1986), The Costs of Protectionism to Developing Countries: An Analysis for Selected Agricultural Products, World Bank Staff Working Paper No. 769, Washington.

MIX
Papier aus verantwortungsvollen Quellen
Paper from responsible sources
FSC® C105338

If you have any concerns about our products,
you can contact us on
ProductSafety@springernature.com

In case Publisher is established outside the EU,
the EU authorized representative is:
Springer Nature Customer Service Center GmbH
Europaplatz 3, 69115 Heidelberg, Germany

Printed by Libri Plureos GmbH
in Hamburg, Germany